Digital Electronic Technology

Digital Electronic Technology

D C Green
M Tech, CEng, MIERE
Formerly Senior Lecturer in Telecommunication Engineering
Willesden College of Technology

Second edition

Longman Group UK Limited,
Longman House, Burnt Mill, Harlow,
Essex CM20 2JE, England
and Associated Companies throughout the world.

First published in Great Britain 1982 by Pitman Books Ltd
Second impression published by Longman Scientific & Technical 1987
Second edition 1988

British Library Cataloguing in Publication Data
Green, D.C.
Digital electronic technology.—2nd ed.
1. Digital electronics
I. Title
621.3815 TK7868.D5

ISBN 0-582-98816-0

Produced by Longman Group (FE) Limited
Printed in Hong Kong

Contents

Preface

This book provides a comprehensive coverage of the more important circuits and techniques used in modern digital electronic circuitry. The varius logic families are discussed and their relative merits considered but the description of the digital circuits, such as counters, is illustrated by reference to the two most popular families, i.e. ttl and cmos, only.

The amount of desirable knowledge for an electronic/telecommunication technician or engineer is very large and greater than could possibly be tackled in a standard 60 hour unit. This means, of course, that some selection of the material to be covered by a particular course is necessary. The UK Business and Technician Education Council (BTEC) has produced a large unit for Level IV Electronics that covers most aspects of both analogue and digital electronics from which colleges are invited to select material to make up one or more 60 hour units.

This book has been written to cover most of the material specified in the digital sections of this large unit. The analogue sections are covered in the companion volume *Electronics* IV.

This book has been written on the assumption that the reader will already possess a knowledge of Electronics and of Digital Electronics up to the standard reached by the level III units of the BTEC. The book should be useful to all higher technician students of digital electronics.

D.C.G.

1 Combinational Logic

Introduction

The *electronic gate* is a circuit that is able to operate on a number of binary signals in order to perform a particular logical function. The types of gate available are the NOT (or inverter), AND, OR, NAND, NOR, exclusive-OR, and the exclusive-NOR (or the coincidence gate). Except for the exclusive-NOR gate, they can be obtained in monolithic integrated circuit form and practical examples will be discussed in the next chapter. The British Standard symbols for these gates are given in Fig. 1.1(*a*). Note that the & sign for the

Fig. 1.1 Gate symbols

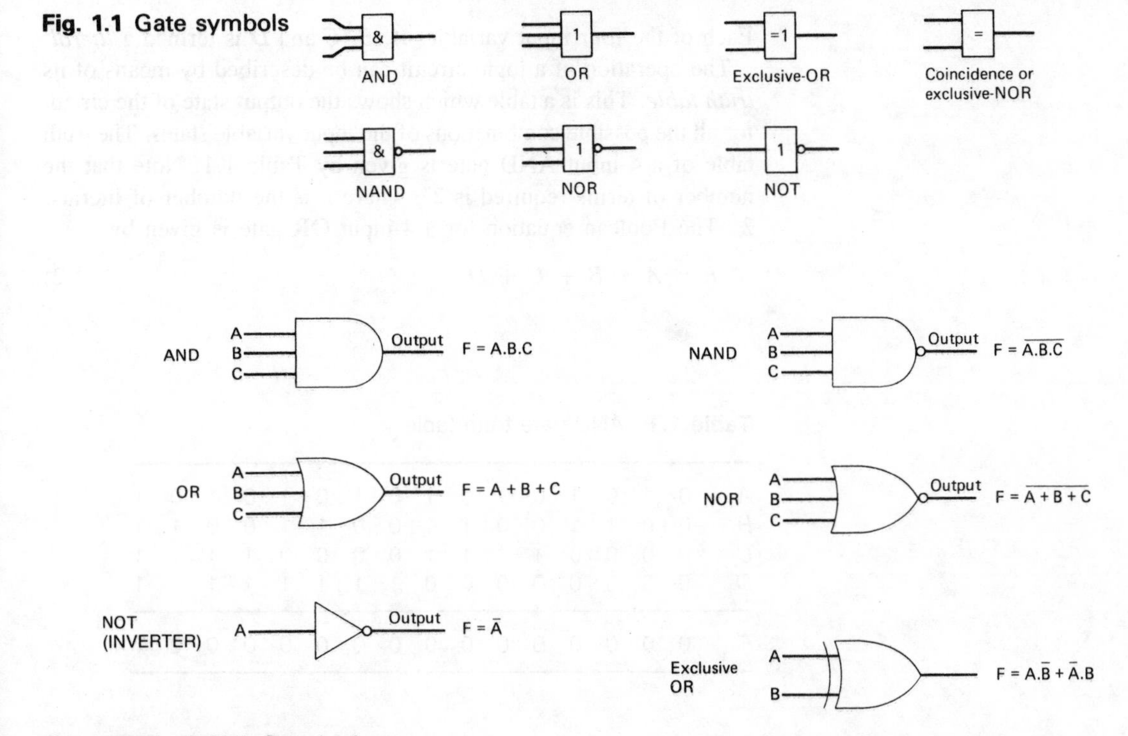

Fig. 1.1(a) British Standard gate symbols (b) Military and American Standard gate symbols

AND and the NAND gates can be replaced by a number indicating the number of inputs that must go high to make the output go high or low. The Military and American Standard sysmbols are given in Fig. 1.1(*b*).

Many logical functions can also be implemented by means of either a medium-scale integrated circuit device known as a *multiplexer* or a *read only memory* (ROM) or a large-scale integrated circuit known as a *programmable logic array*.

The Basic Gates

1 The **AND** gate is a logic element having two or more input terminals and only one output terminal. Its output is at logical 1 *only* when *all* of its inputs are at logical 1. If any one or more of the inputs are at logic 0, the output of the gate will also be at 0. The output F of a 4-input AND gate can be expressed using Boolean algebra as

$$F = A.B.C.D \tag{1.1}$$

The symbol for the AND logical function is the dot as shown but the dot is usually omitted, and the AND function is expressed as

$$F = A\ B\ C\ D$$

Each of the four input variables A, B, C and D is termed a *literal*.

The operation of a logic circuit can be described by means of its *truth table*. This is a table which shows the output state of the circuit for all the possible combinations of the input variable states. The truth table of a 4-input AND gate is given by Table 1.1. Note that the number of terms required is 2^n, where n is the number of literals.

2 The Boolean equation for a 4-input **OR** gate is given by

$$F = A + B + C + D \tag{1.2}$$

Table 1.1 AND gate truth table

A	0	1	0	1	0	1	0	1	0	1	0	1	0	1	0	1
B	0	0	1	1	0	0	1	1	0	0	1	1	0	0	1	1
C	0	0	0	0	1	1	1	1	0	0	0	0	1	1	1	1
D	0	0	0	0	0	0	0	0	1	1	1	1	1	1	1	1
F	0	0	0	0	0	0	0	0	0	0	0	0	0	0	0	1

Table 1.2 OR gate truth table

A	0	1	0	1	0	1	0	1	0	1	0	1	0	1	0	1
B	0	0	1	1	0	0	1	1	0	0	1	1	0	0	1	1
C	0	0	0	0	1	1	1	1	0	0	0	0	1	1	1	1
D	0	0	0	0	0	0	0	0	1	1	1	1	1	1	1	1
F	0	1	1	1	1	1	1	1	1	1	1	1	1	1	1	1

The + symbol is the Boolean symbol for the OR logical function. Its truth table is given in Table 1.2. It is clear that the output of an OR gate is at logic 1 whenever any one, or more, of its inputs are in the logical 1 state.

3 The **NAND** logical function is equivalent to the AND followed by an inversion, and hence the output of a NAND gate is at logic 0 *only* when *all* of its inputs are at 1. Otherwise its output is 1. This action is illustrated by the Boolean equation for a 4-input NAND gate (eqn. (1.3)) and by the truth table (Table 1.3).

$$F = \overline{A\ B\ C\ D} \qquad (1.3)$$

Table 1.3 NAND gate truth table

A	0	1	0	1	0	1	0	1	0	1	0	1	0	1	0	1
B	0	0	1	1	0	0	1	1	0	0	1	1	0	0	1	1
C	0	0	0	0	1	1	1	1	0	0	0	0	1	1	1	1
D	0	0	0	0	0	0	0	0	1	1	1	1	1	1	1	1
F	1	1	1	1	1	1	1	1	1	1	1	1	1	1	1	0

Table 1.4 NOR gate truth table

A	0	1	0	1	0	1	0	1	0	1	0	1	0	1	0	1
B	0	0	1	1	0	0	1	1	0	0	1	1	0	0	1	1
C	0	0	0	0	1	1	1	1	0	0	0	0	1	1	1	1
D	0	0	0	0	0	0	0	0	1	1	1	1	1	1	1	1
F	1	0	0	0	0	0	0	0	0	0	0	0	0	0	0	0

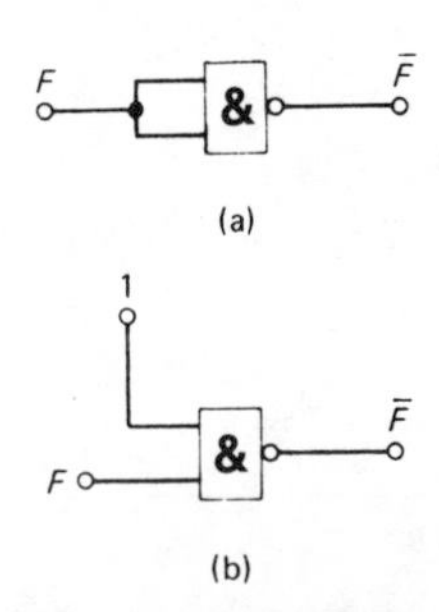

Fig. 1.2 The NAND gate as an inverter

A NAND gate can be used to provide the NOT, or inversion, function by connecting all of its input terminals together (Fig. 1.2(*a*)) or by connecting all but one of its inputs to logic 1 (Fig. 1.2(*b*)). The NAND gate is one of the most commonly employed small-scale integrated circuits and it is frequently used to produce both the AND and the OR functions. This is demonstrated later (p. 19).

4 The other commonly employed gate is the **NOR** gate. The truth table of a 4-input NOR gate is given by Table 1.4. This table should make it clear that the NOR gate gives the NOT OR logic operation. This means that its output *F* is at logical 1 only when *all* of its inputs are at logical 0. The NOR gate can also be used to produce the AND and the OR logic functions and can be used as an inverter (see Figs 1.3(*a*) and (*b*)). Either all of its input terminals are connected together or all but one of its inputs are connected to logic 0.

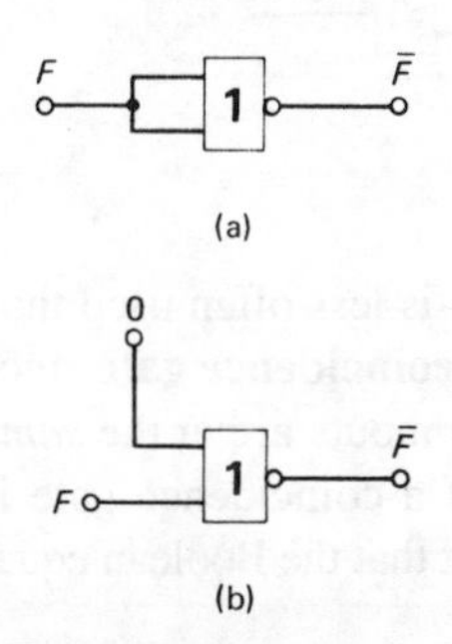

Fig. 1.3 The NOR gate as an inverter

An alternative to the use of a NAND or a NOR gate as an inverter is to use a hex (six) inverter integrated circuit, e.g. the ttl 7404.

5 The **exclusive-OR** gate has two input terminals and performs the logical function

$$F = A\bar{B} + \bar{A}B \tag{1.4}$$

Table 1.5 Exclusive-OR truth table

A	0	1	0	1
B	0	0	1	1
F	0	1	1	0

The truth table describing this function is given by Table 1.5. The output of the gate is at logical 1 only when either one but *not* both of its inputs is at logical 1. If both inputs are at 0, *or* at 1, the output of the gate will be at logical 0.

The exclusive-OR gate can be purchased as an integrated circuit package or can be made up from a suitable combination of AND, NOT, and OR gates (Fig. 1.4(*a*)) or by using NAND gates only (Fig. 1.4(*b*)).

Commercially available ic packages, e.g. ttl 7486, only include 2-input exclusive-OR gates. If the functions

$$F = A \oplus B \oplus C \quad \text{or} \quad F = A \oplus B \oplus C \oplus D$$

are required, then two or three exclusive-OR gates can be connected in the manner shown by Fig. 1.4(*c*). (The symbol $\oplus$ denotes the exclusive-OR function.)

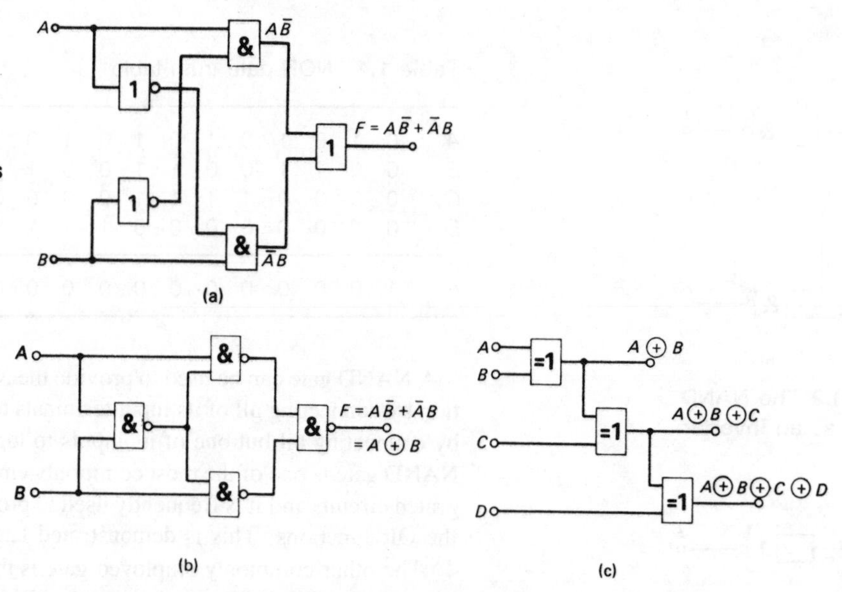

Fig 1.4 Exclusive-OR gate (*a*) using AND, OR, and NOT gates, (*b*) using NAND gates only, (*c*) for three or four literals

Table 1.6 Exclusive-NOR truth table

A	0	1	0	1
B	0	0	1	1
F	1	0	0	1

6 The last type of gate, the **exclusive-NOR**, is less often used than any of the other types. It is also known as the **coincidence gate** since its output is at logical 1 *only* when *both* of its inputs are at the *same* logical state, be it 0 or 1. The truth table of a coincidence gate is given by Table 1.6. From this table it is evident that the Boolean equation describing the coincidence gate is

$$F = \bar{A}\bar{B} + AB \tag{1.5}$$

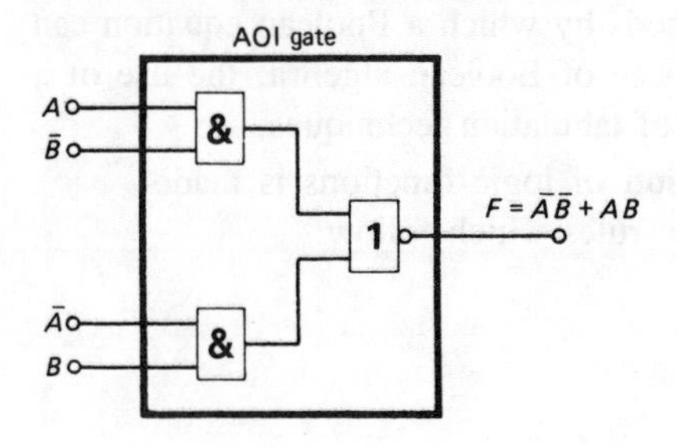

Fig. 1.5 Exclusive-NOR function generated by an AOI gate

It should also be observed that the output F of the gate is always the inverse of the output of the exclusive-OR gate. It is for this reason that the circuit is also known as the exclusive-NOR gate. The exclusive-NOR gate can be achieved in three different ways:

1) directly, using AND, NOT, and OR gates
2) by inverting the output of an exclusive-OR gate
3) using an *AOI gate* (p. 34) as shown by Fig. 1.5.

Boolean Equations

In the design of a combinational logic circuit, the truth table of the required operation can be written down and then used to derive the Boolean equation that expresses the output of the circuit in terms of the literals. The equation thus obtained can generally be simplified to reduce the number of gates required for its implementation. Very often the implementation is carried out using either NAND or NOR gates exclusively.

Boolean equations are often in either the *sum-of-products* form, e.g.

$$F = A\bar{B} + \bar{A}B$$

or the *product-of-sums* form, e.g.

$$F = (A + \bar{B})(\bar{A} + B)$$

Each of the product terms in a sum-of-products equation is known as a *minterm*, while each sum term in a product-of-sums equation is called a *maxterm*.

In fact, any Boolean equation can be expressed in either of these forms, and one or the other may be the simpler to implement in a particular case.

It will be seen (rule **6** on p. 6) that if any minterm in a sum-of-products equation is equal to logical 1 then the sum F of that equation is also 1. Conversely (rule **8** on p. 6), if any maxterm in a product-of-sums equation is at logical 0 then the product F will also be at 0.

The Simplification of Boolean Equations

When the Boolean equation describing the logic operation of a circuit has been obtained, an attempt is often made to simplify, or **minimize**, the equation.

An equation is said to be minimized when it *a*) contains the lowest possible number of literals and *b*) the lowest possible number of terms. The minimized equation will usually require the fewest number of gates possible for its implementation, although this may not necessarily also give the least number of ic packages. Further, the minimal solution may well be subject to *race-hazards* (p. 27).

There are three main methods by which a Boolean equation can be simplified; these are the use of Boolean algebra, the use of a Karnaugh map, and the use of tabulation techniques.

The **algebraic simplification** of logic functions is made easier by the use of the logic rules which follow:

1 $A + \bar{A} = 1$
2 $A + A = A$
3 $AA = A$
4 $A\bar{A} = 0$
5 $A + 0 = A$
6 $A + 1 = 1$
7 $A.1 = A$
8 $A.0 = 0$
9 $AB = BA$ } Commutative law
10 $A + B = B + A$ }
11 $A(B + C) = AB + AC$ } Distributive law
12 $A + (BC) = (A + B)(A + C)$ }
13 $A + B + C = (A + B) + C = A + (B + C)$ } Associative law
14 $ABC = A(BC) = (AB)C$ }
15 $A(B + \bar{B}) = A$
16 $A + AB = A$
17 $A(A + B) = A$
18 $A + \bar{A}B = A + B$
19 $B(A + \bar{B}) = AB$
20 $(A + B)(B + C)(C + \bar{A}) = (A + B)(C + \bar{A})$
21 $\overline{A + B} = \bar{A}\bar{B}$
22 $\overline{AB} = \bar{A} + \bar{B}$
23 $AB + BC + \bar{A}C = AB + \bar{A}C$

21 and 22 are known as **De Morgan's rules**. The accuracy of any of these rules can be confirmed with the aid of a truth table.

Example 1.1

Simplify $F = \bar{A}(B + \bar{C})(A + \bar{B} + C)\bar{A}\bar{B}\bar{C}$

Solution

Multiplying out gives

$$F = (AB + B\bar{B} + BC + A\bar{C} + \bar{B}\bar{C} + C\bar{C})\bar{A}\bar{A}\bar{B}\bar{C}$$

Rules 3 and 4 give

$$F = (AB + BC + A\bar{C} + \bar{B}\bar{C})\bar{A}\bar{B}\bar{C}$$

Multiplying out again and applying rules 3 and 4 gives

$$F = \bar{A}\bar{B}\bar{C} \quad (Ans)$$

Example 1.2

Simplify $F = (A + B)(\overline{AB} + C) + AB$

Solution

Rule 22 gives

$$F = (A + B)(\bar{A} + \bar{B} + C) + AB$$

Multiplying out and using rule 4 gives

$$\begin{aligned} F &= A\bar{B} + AC + \bar{A}B + BC + AB \\ &= A(B + \bar{B}) + AC + \bar{A}B + BC \\ &= A(1 + C) + \bar{A}B + BC && \text{from rule 1} \\ &= A + \bar{A}B + BC && \text{from rule 6} \\ &= A + B + BC && \text{from rule 18} \\ &= A + B(1 + C) && \text{from rule 11} \\ &= A + B \quad (Ans) && \text{from rule 6} \end{aligned}$$

With some practice it is possible to write several of the above steps down at once.

Example 1.3

Simplify the equation $F = ABC + ABD + \bar{A}B\bar{C} + CD + B\bar{D}$

Solution

$$\begin{aligned} F &= ABC + ABD + \bar{A}B\bar{C} + CD + B\bar{D} \\ &= ABC + \bar{A}B\bar{C} + CD + B(\bar{D} + AD) \\ &= ABC + \bar{A}B\bar{C} + CD + B(\bar{D} + A) && \text{from rule 18} \\ &= AB(1 + C) + \bar{A}B\bar{C} + CD + B\bar{D} \\ &= AB + \bar{A}B\bar{C} + CD + B\bar{D} && \text{from rule 6} \\ &= B(A + \bar{A}\bar{C}) + CD + B\bar{D} \\ &= AB + B\bar{C} + CD + B\bar{D} && \text{from rule 18} \\ &= AB + B\bar{C} + CD + BC + B\bar{D} && \text{from rule 23} \\ &= AB + B(C + \bar{C}) + CD + B\bar{D} \\ &= AB + B + CD + B\bar{D} && \text{from rule 1} \\ &= B(1 + A + \bar{D}) + CD \\ &= B + CD \quad (Ans) && \text{from rule 6} \end{aligned}$$

Conversion of a Sum-of-Products Equation into the Equivalent Product-of-Sums Form

The conversion of an equation in product-of-sums form into the equivalent sum-of-products form is easily accomplished by merely multiplying out (see Examples 1.1 and 1.2). The reverse process, namely converting an equation from its sum-of-products form into its equivalent product-of-sums form, is more difficult. Consider the equation

$$F = ABC + \bar{A}B\bar{C}$$

The first step is to obtain the complement of the equation

$$
\begin{aligned}
\bar{F} &= \overline{ABC + \bar{A}B\bar{C}} \\
&= \overline{ABC} \bullet \overline{\bar{A}B\bar{C}} \\
&= (\bar{A} + \bar{B} + \bar{C})(A + \bar{B} + C)
\end{aligned}
$$

Multiplying out and then simplifying gives

$$
\begin{aligned}
\bar{F} &= \bar{A}\bar{B} + \bar{A}C + A\bar{B} + \bar{B} + \bar{B}C + A\bar{C} + \bar{B}\bar{C} \\
&= \bar{B}(1 + \bar{A} + A + C + \bar{C}) + \bar{A}C + A\bar{C} \\
&= \bar{B} + \bar{A}C + A\bar{C}
\end{aligned}
$$

Applying De Morgan's rule again:

$$F = \overline{\bar{B} + \bar{A}C + A\bar{C}} = B(\overline{\bar{A}C})(\overline{A\bar{C}})$$

$$\text{or} \quad F = B(A + \bar{C})(\bar{A} + C)$$

The algebraic method of simplying Boolean equations possesses the disadvantages that *a*) it can be very time-consuming and *b*) it requires considerable practice and experience before the most appropriate approach and/or rule can be selected quickly. Usually it is better to employ the mapping method described later.

Canonical Form

The **canonical form** of a Boolean equation is one in which *each* term contains *all* of the literals *once* only. If two equations are written down in their canonical forms, they can be compared term by term and this fact will be utilized later when a tabular method of simplifying Boolean equations is described. Also, the canonical form of a sum-of-products equation is required when the equation is to be implemented using a ROM (p. 95) or a multiplexer (p. 30).

Example 1.4

Write the equation $\bar{A}B + \bar{C}B$ in its canonical form.

Solution

$$
\begin{aligned}
F &= \bar{A}B(C + \bar{C}) + \bar{C}B(A + \bar{A}) \\
&= \bar{A}BC + \bar{A}B\bar{C} + AB\bar{C} + \bar{A}B\bar{C} \\
&= \bar{A}BC + \bar{A}B\bar{C} + AB\bar{C} \quad (Ans)
\end{aligned}
$$

Example 1.5

Write the equation $(\bar{A} + B)(\bar{C} + B)$ in its canonical form.

Solution

$$
\begin{aligned}
F &= (\bar{A} + B + C)(\bar{A} + B + \bar{C})(A + B + \bar{C})(\bar{A} + B + C) \\
&= (\bar{A} + B + C)(\bar{A} + B + \bar{C})(A + B + \bar{C}) \quad (Ans)
\end{aligned}
$$

Duality

Every sum-of-products equation has a dual product-of-sums equation and vice versa, and this fact provides another method which can be used in the simplification of Boolean equations.

The **dual** of an equation is simply obtained by merely replacing every AND symbol (.) by the OR symbol (+) and vice versa. For example,

the dual of $F = (\bar{A} + B + C)(A + D)$ is $F' = \bar{A}BC + AD$

Example 1.6

Use duality to simplify Example 1.2.

Solution

$$
\begin{aligned}
F &= (A + B)(\bar{A} + \bar{B} + C) + AB \\
F' &= (AB + \bar{A}\bar{B}C)(A + B) \\
&= AB
\end{aligned}
$$

Therefore $F = A + B$ *(Ans)*

Clearly in this case the algebra involved is much simpler than previously.

The complement $\bar{F}$ of a Boolean equation can be obtained by replacing each literal by its complement in the corresponding dual equation.

Thus if $F = AB + C$, then $F' = (A + B)C$ and so

$$\bar{F} = (\bar{A} + \bar{B})\bar{C} = \bar{A}\bar{C} + \bar{B}\bar{C}$$

Example 1.7

Use duality to find the complement of the equation

$$F = ABC + \bar{A}D$$

Check your answer using De Morgan's rules.

Solution

$$
\begin{aligned}
F' &= (A + B + C)(\bar{A} + D) \\
\text{so } \bar{F} &= (\bar{A} + \bar{B} + \bar{C})(A + \bar{D}) \\
&= A\bar{B} + A\bar{C} + \bar{A}\bar{D} + \bar{B}\bar{D} + \bar{C}\bar{D} \quad \textit{(Ans)}
\end{aligned}
$$

Using De Morgan

$$
\begin{aligned}
\bar{F} &= \overline{ABC + \bar{A}D} \\
&= \overline{ABC}.\overline{\bar{A}D} = (\bar{A} + \bar{B} + \bar{C})(A + \bar{D}) \quad \text{as before}
\end{aligned}
$$

Example 1.8

Find the complement of $F = ABC + \bar{A}B\bar{C} + \bar{A}\bar{B}C$: *a*) directly, *b*) using duality.

Solution

a) $\bar{F} = \overline{ABC + \bar{A}B\bar{C} + \bar{A}\bar{B}C} = (\overline{ABC})(\overline{\bar{A}B\bar{C}})(\overline{\bar{A}\bar{B}C})$
$= (\bar{A} + \bar{B} + \bar{C})(A + \bar{B} + C)(A + B + \bar{C})$
$= (\bar{A}\bar{B} + \bar{A}C + A\bar{B} + \bar{B} + \bar{B}\bar{C} + A\bar{C} + \bar{B}C)(A + B + \bar{C})$
$= (\bar{A}C + A\bar{C} + \bar{B})(A + B + \bar{C})$
$= \bar{A}BC + A\bar{C} + A\bar{B} + \bar{B}\bar{C}$ (*Ans*)

b) $F' = (A + B + C)(\bar{A} + B + \bar{C})(\bar{A} + \bar{B} + C)$
$\bar{F} = (\bar{A} + \bar{B} + \bar{C})(A + \bar{B} + C)(A + B + \bar{C})$ as before

The Karnaugh Map

The Karnaugh map provides a convenient method of simplifying sum-of-products Boolean equations in which the function to be simplified is displayed diagrammatically on a map of squares. Each square maps one term of the function. The number of squares is equal to 2^n, where n is the number of literals in the function. Thus, if the equation to be simplified has three literals A, B and C, then $n = 3$ and $2^3 = 8$, and so 8 squares are needed. The rows and columns of the Karnaugh map can be labelled as shown for 4, 8 and 16 square maps. (The labels *in* the squares are not normally given.)

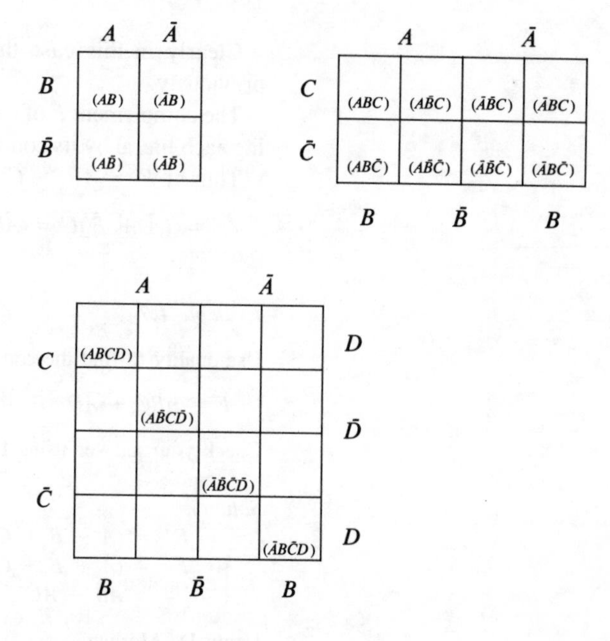

Often, a Karnaugh map is labelled in a slightly different manner to that shown above. The alternative labelling is shown below for a 16-square map.

The number 1 written in a square indicates the presence, in the function being mapped, of the term represented by that square. The number 0 written in a square means that the particular term is not present in the mapped function.

To simplify an equation using the Karnaugh map adjacent squares

CD \ AB	00	01	11	10
00	$(\bar{A}\bar{B}\bar{C}\bar{D})$			
01		$(\bar{A}B\bar{C}D)$		
11			$(ABCD)$	
10				$(A\bar{B}C\bar{D})$

containing 1 are looped together. This step eliminates any terms of the form $A\bar{A}$.

In this context, *adjacent* means *a*) side-by-side in the horizontal and in the vertical directions (but *not* diagonal), *b*) the right-hand and left-hand sides, and the top and bottom, of the map and *c*) the *four* corner squares. Squares can be looped together in two, fours or eights *only*.

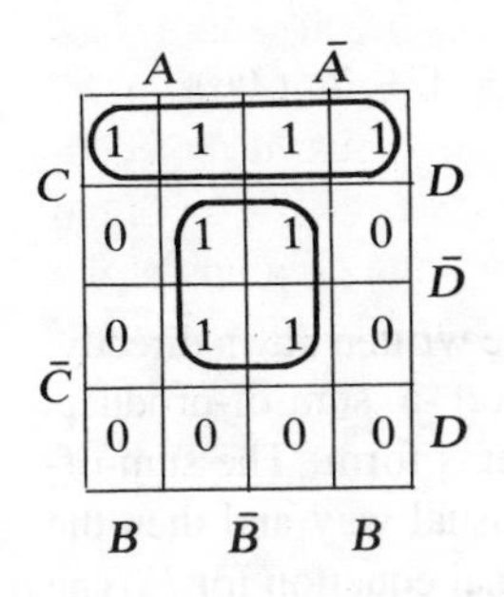

Example 1.9

Use a Karnaugh map to simplify the equation

$$F = ACD + \bar{A}BCD + \bar{B}\bar{C}\bar{D} + \bar{A}\bar{B}C\bar{D} + \bar{A}\bar{B}CD + A\bar{B}C\bar{D}$$

Solution

The mapping of the equation is shown.

Looping the adjacent squares in two groups of four simplifies the equation to

$$F = CD + \bar{B}\bar{D} \quad (Ans)$$

When $\bar{F}$ is required it will probably be quicker to loop the 0 squares.

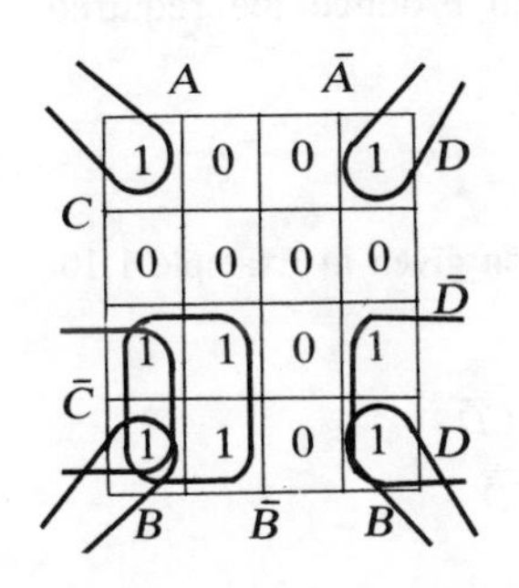

Example 1.10

Simplify the equation $F = ABCD + \bar{A}BCD + A\bar{C}D + A\bar{C}\bar{D} + \bar{A}B\bar{C}$ and obtain $\bar{F}$.

Solution

The mapping of F is shown.

1 Looping the 1 squares, $F = BD + A\bar{C} + B\bar{C}$.
Therefore,

$$\begin{aligned} \bar{F} &= \overline{BD + A\bar{C} + B\bar{C}} \\ &= \overline{BD}\,\overline{A\bar{C}}\,\overline{B\bar{C}} = (\bar{B} + \bar{D})(\bar{A} + C)(\bar{B} + C) \\ &= (\bar{A}\bar{B} + \bar{B}C + \bar{A}\bar{D} + C\bar{D})(\bar{B} + C) \\ &= \bar{A}\bar{B} + \bar{B}C + \bar{A}\bar{B}\bar{D} + \bar{B}C\bar{D} + \bar{A}\bar{B}C + \bar{B}C + \bar{A}C\bar{D} + C\bar{D} \\ &= \bar{A}\bar{B} + \bar{B}C + C\bar{D} \quad (Ans) \end{aligned}$$

2 Looping the 0 squares

$$\bar{F} = \bar{A}\bar{B} + \bar{B}C + C\bar{D} \quad \text{(directly)} \quad (Ans)$$

The Karnaugh map can also be used to simplify an equation of the product-of-sums form. Either the equation can be multiplied out into the equivalent sum-of-product form, or each term can be mapped separately and then the individual maps can be combined by ANDing them. Note that for corresponding squares in each map

$$1\;1 = 1 \qquad 1\;0 = 0 \qquad 0\;0 = 0$$

Example 1.11

Repeat Example 1.2 using a Karnaugh map.

Solution

Applying De Morgan's rule, $F = (A + B)(\bar{A} + \bar{B} + C) + AB$.

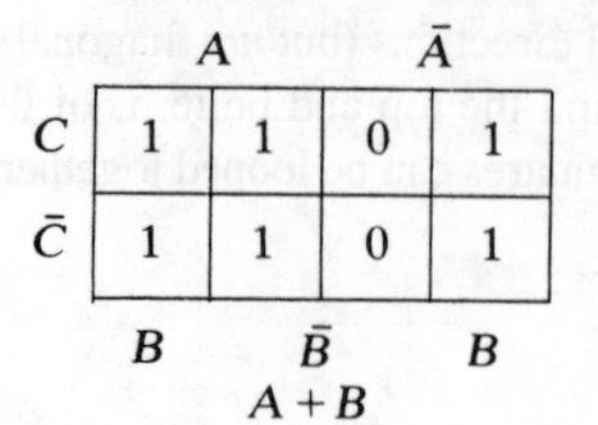

A+B

	A	A	Ā	Ā
C	1	1	1	1
$\bar{C}$	0	1	1	1
	B	$\bar{B}$	$\bar{B}$	B

$\bar{A}+\bar{B}+C$

	A	A	Ā	Ā
C	1	1	0	1
$\bar{C}$	0	1	0	1
	B	$\bar{B}$	$\bar{B}$	B

$(A+B)(\bar{A}+\bar{B}+C)$

Lastly, mapping AB

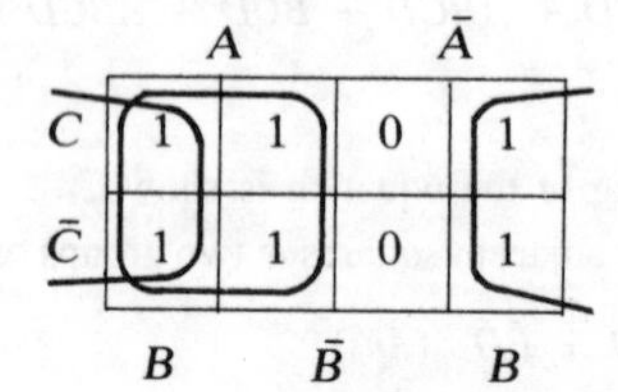

From the mapped squares, $F = A + B$ (*Ans*)

With a little practice the final mapping can be written down directly.

A Karnaugh map can be used to convert a sum-of-products equation into its corresponding product-of-sums form. The sum-of-products equation should be mapped in the usual way and then the 0 entries should be looped to obtain the minimal equation for $\bar{F}$. The application of De Morgan's rule will then produce the required product-of-sums equation.

Example 1.12

Obtain the product-of-sums form of the equation given in Example 1.10.

Solution

From the map on p. 11 $\quad \bar{F} = \bar{A}\bar{B} + \bar{B}C + C\bar{D}$

Therefore $F = \overline{\bar{A}\bar{B} + \bar{B}C + C\bar{D}} = \overline{\bar{A}\bar{B}}.\overline{\bar{B}C}.\overline{C\bar{D}}$

or $F = (A + B)(B + \bar{C})(\bar{C} + D)$ (*Ans*)

The use of De Morgan can be avoided and the required equation taken directly from the map if, for each grouping of 0 squares, the complement of the map labelling is used.

Thus

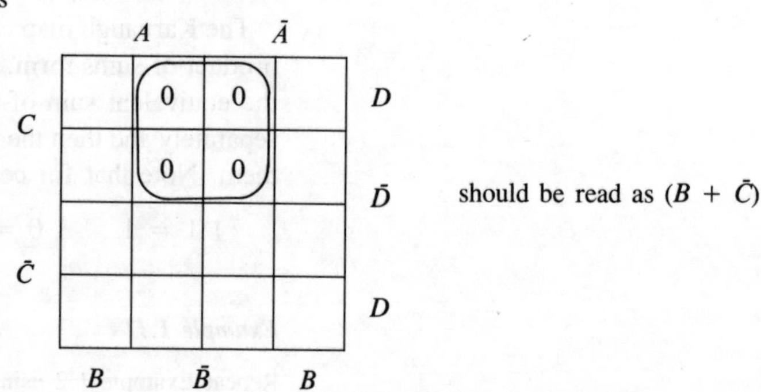

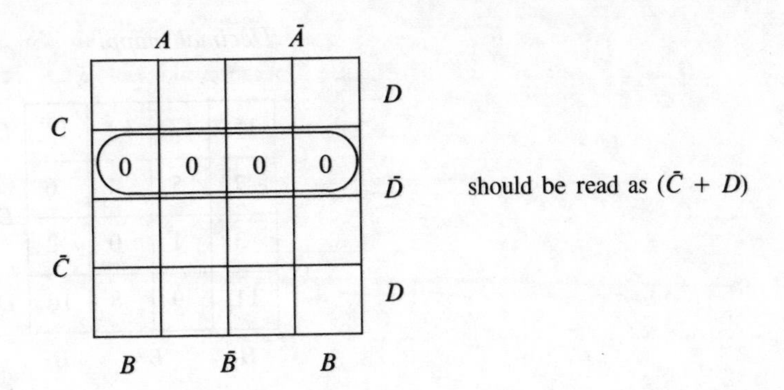

should be read as $(\bar{C} + D)$

and so on.

Example 1.13

Obtain the minimal product-of-sums form of

$$F = AB + \bar{A}CD + \bar{A}\bar{B}C\bar{D} + \bar{A}BC\bar{D}$$

Solution

The mapping is

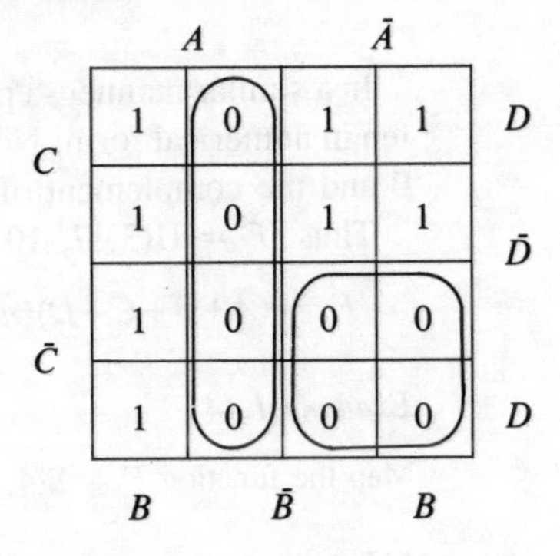

Looping the 0 squares gives

$$F = (\bar{A} + B)(A + C) \quad (Ans)$$

Decimal Representation of Terms

It is sometimes convenient to express the various terms of a Boolean equation in their decimal form. Thus, referring to Table 1.7, assuming A to be the least significant bit, the decimal mapping is as shown.

The decimal form of a sum-of-products Boolean equation can be written as

$$F = \Sigma(ABCD)$$

Thus, $F = \Sigma(3, 7, 10, 12)$ represents the equation

$$F = AB\bar{C}\bar{D} + ABC\bar{D} + \bar{A}B\bar{C}D + \bar{A}\bar{B}CD$$

Decimal mapping

	A		$\bar{A}$		
C	15	13	12	14	D
	7	5	4	6	$\bar{D}$
$\bar{C}$	3	1	0	2	
	11	9	8	10	D
	B	$\bar{B}$		B	

$CD \backslash AB$	00	01	11	10
00	0	2	3	1
01	8	10	11	9
11	12	14	15	13
10	4	6	7	5

Table 1.7 Decimal representation of Boolean equations

Decimal number	0	1	2	3	4	5	6	7	8	9	10	11	12	13	14	15
A	0	1	0	1	0	1	0	1	0	1	0	1	0	1	0	1
B	0	0	1	1	0	0	1	1	0	0	1	1	0	0	1	1
C	0	0	0	0	1	1	1	1	0	0	0	0	1	1	1	1
D	0	0	0	0	0	0	0	0	1	1	1	1	1	1	1	1

In a similar manner, a products-of-sums equation can also be written in numerical form. Now, however, each literal is given the value 0 and the complement of each literal has the value 1.

Thus, $F = \Pi(3, 7, 10, 12)$ represents the equation

$$F = (\bar{A}+\bar{B}+C+D)(\bar{A}+\bar{B}+\bar{C}+D)(A+\bar{B}+C+\bar{D})(A+B+\bar{C}+\bar{D})$$

Example 1.14

Map the function $F = \Sigma(4, 6, 10, 11, 14, 15)$ and then simplify.

Solution

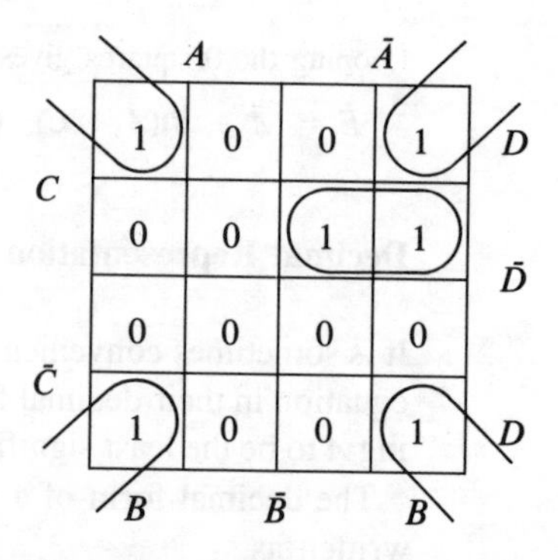

From the map $F = BD + \bar{A}C\bar{D}$ *(Ans)*

Example 1.15

Map the function $F = \Pi(4, 6, 10, 11, 14, 15)$ and simplify.

Solution

	A	A	$\bar{A}$	$\bar{A}$	
C	0	1	1	0	D
C	1	1	0	0	$\bar{D}$
$\bar{C}$	1	1	1	1	$\bar{D}$
$\bar{C}$	0	1	1	0	D
	B	$\bar{B}$	$\bar{B}$	B	

From the map $F = (\bar{B} + \bar{D})(A + \bar{C} + D)$ (*Ans*)

"Don't-Care" Conditions

The truth tables of some logic circuits contain certain combinations of literals for which the output F is unimportant and so they can be either 1 or 0. Such combinations are said to be "don't-care" conditions or states. When a Boolean function is mapped, any don't-care terms are represented by an X and can be looped in with *either* the 1 squares *or* the 0 squares in the simplification of the function. An X square should be looped with a group of 1 squares if the looping then gives a greater reduction in the plotted equation. Suppose, for example, that the mapping of a particular function is

	A	A	$\bar{A}$	$\bar{A}$	
C	1	1	0	X	D
C	1	X	0	1	$\bar{D}$
$\bar{C}$	X	1	1	1	$\bar{D}$
$\bar{C}$	1	1	0	1	D
	B	$\bar{B}$	$\bar{B}$	B	

Looping the 1 squares only gives

$$F = AD + ABC + \bar{A}B\bar{D} + \bar{A}B\bar{C} + \bar{B}\bar{C}\bar{D}$$

If the don't-care squares are looped together with the 1 squares, the function can be reduced to the much simpler result:

$$F = A + B + \bar{C}\bar{D}$$

Karnaugh Map for More than Four Variables

The number of squares in a Karnaugh map must be equal to 2^n where n is the number of literals. This means that five literals A, B, C, D and E will require $2^5 = 32$ squares. These can be obtained by drawing two 16-square maps side-by-side, one of which represents E while the other represents $\bar{E}$.

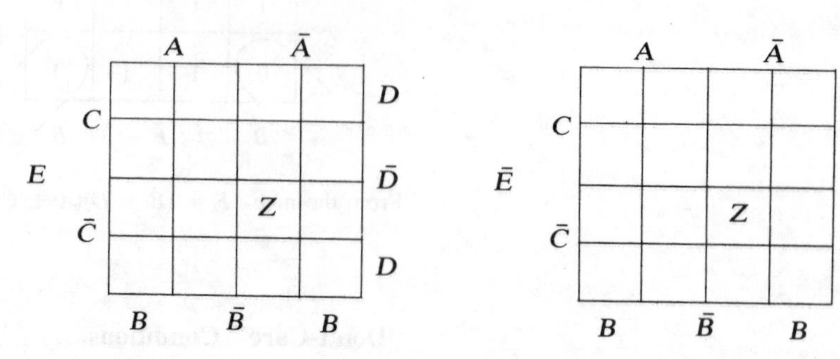

The 32-square map is used to simplify a 5-literal Boolean equation in similar manner to that previously described. Note the similarly positioned squares in each map, e.g. the squares marked Z are considered to be adjacent for looping purposes.

Example 1.16

Simplify the Boolean equation

$$F = \Sigma(2, 5, 6, 9, 10, 14, 15, 16, 18, 20, 22, 26, 29, 30, 31).$$

Solution

The mapping is E

E	A, B	A, $\bar{B}$	$\bar{A}$, $\bar{B}$	$\bar{A}$, B	
C	1 (15)	0 (13)	0 (12)	1 (14)	D
C	0 (7)	1 (5)	0 (4)	1 (6)	$\bar{D}$
$\bar{C}$	0 (3)	0 (1)	0 (0)	1 (2)	$\bar{D}$
$\bar{C}$	0 (11)	1 (9)	0 (8)	1 (10)	D

$\bar{E}$	A, B	A, $\bar{B}$	$\bar{A}$, $\bar{B}$	$\bar{A}$, B	
C	1 (31)	1 (29)	0 (28)	1 (30)	D
C	0 (23)	0 (21)	1 (20)	1 (22)	$\bar{D}$
$\bar{C}$	0 (19)	0 (17)	1 (16)	1 (18)	$\bar{D}$
$\bar{C}$	0 (27)	0 (25)	0 (24)	1 (26)	D

From the map:

a) Squares 5 and 9 cannot be looped: $A\bar{B}C\bar{D}E + A\bar{B}\bar{C}DE$

b) Squares 29 and 31 can be looped: $ACD\bar{E}$

c) Squares 14, 15, 30 and 31 can be looped: BCD

d) Squares 16, 18, 20 and 22 can be looped: $\bar{A}\bar{D}\bar{E}$

e) Squares 2, 6, 10, 14, 18, 22, 26 and 30 can be looped: $\bar{A}B$

Therefore,

$$F = A\bar{B}CD\bar{E} + A\bar{B}\bar{C}DE + ACD\bar{E} + BCD + \bar{A}\bar{D}\bar{E} + \bar{A}B \quad (Ans)$$

Clearly, the use of the Karnaugh map will become increasingly cumbersome if there are more than five literals and in such cases it is better to employ a tabular method of minimization.

Tabular Simplification of Boolean Equations

The Karnaugh map provides a convenient method for the simplification of Boolean equations with up to four literals and can be used with five literals. When the number of literals is in excess of this number, it is generally easier to employ a tabular method of simplification. The method to be described is due to Quine and McCluskey.

a) Write down the equation to be simplified in its *canonical* form.

b) Write all the canonical terms in a column (column A), starting with those terms that include the *most complemented* literals.

c) In column B write, for each term in *b*), a 1 for an uncomplemented literal and 0 for a complemented literal.

d) Divide column B into *groups* containing, in order, no 1s, a single 1, two 1s, three 1s, and so on.

e) An attempt is now made to ''match'' each term within a group with another group. Here *matching* means that the two terms differ from one another by only *one* literal. For example, the terms 0001 and 1001 would match. Each term that is matched with another term is ticked. A term should be matched as many times as possible.

f) The *matched terms* are then entered in column C but the two different literals have cancelled out and are hence omitted.

g) When column C has been completed, it is divided into further *groups* as before and another set of matched terms is found.

h) These matched terms are entered in column D and so on.

i) When no more matchings can be found, the table is inspected for all the *unticked* terms. These are known as the PRIME IMPLICANTS. This may be the simplest form of the equations but it is possible that some further simplification can be achieved.

j) A *prime implicant table* is now constructed which is used to locate any common terms and thereby allow a further reduction in the function.

The process as described may seem rather lengthy but with some practice it can be done quite quickly and can be programmed on to a computer.

Example 1.17

Solve the equation $F = AB + B\bar{C} + CD + \bar{B}D$ using a tabular method.

Solution

This is, of course, an equation that could easily be simplified using a Karnaugh map.

Step A Writing the given equation in its canonical form:

$$F = AB(C+\bar{C})(D+\bar{D}) + B\bar{C}(A+\bar{A})(D+\bar{D}) + CD(A+\bar{A})(B+\bar{B}) + \bar{B}D(A+\bar{A})(C+\bar{C})$$

$$= ABCD + AB\bar{C}D + ABC\bar{D} + AB\bar{C}\bar{D} + \bar{A}B\bar{C}D + \bar{A}B\bar{C}\bar{D} + A\bar{B}CD + \bar{A}BCD + \bar{A}\bar{B}CD + A\bar{B}\bar{C}D + \bar{A}\bar{B}\bar{C}D$$

Step B These terms are now listed in column A (of Table 1.8).

Step C Column B is then produced by writing 1 for each uncomplemented literal and 0 for each complemented literal in column A.

Step D Column B is then divided into 4 groups by drawing horizontal lines. In each group the number of 1s is the same.

Step E The matching process can now begin. Terms 1 and 4 differ only in their second bit from the left and are matched together and ticked. 0-01 is then entered into column C. Term 1 can also be matched with term 5 and so 00-1 is entered into column C.

Table 1.8 Tabular simplification of Boolean equations

	A	B	C	D	E
1	$\bar{A}\bar{B}\bar{C}D$	0 0 0 1 √	0 – 0 1 √	0 – – 1 √	– – – 1
2	$\bar{A}B\bar{C}\bar{D}$	0 1 0 0 √	0 0 – 1 √	– – 0 1 √	– – – 1 } → D
3	$AB\bar{C}\bar{D}$	1 1 0 0 √	– 0 0 1 √	– 0 – 1 √	– – – 1
4	$\bar{A}B\bar{C}D$	0 1 0 1 √	– 1 0 0 √	– 1 0 – ⟷ $B\bar{C}$	
5	$\bar{A}\bar{B}CD$	0 0 1 1 √	0 1 0 – √	1 1 – – ⟷ AB	
6	$A\bar{B}\bar{C}D$	1 0 0 1 √	1 1 0 – √	– 1 – 1 √	
7	$AB\bar{C}D$	1 1 0 1 √	1 1 – 0 √	– – 1 1 √	
8	$ABC\bar{D}$	1 1 1 0 √	– 1 0 1 √	1 – – 1 √	
9	$\bar{A}BCD$	0 1 1 1 √	0 1 – 1 √		
10	$A\bar{B}CD$	1 0 1 1 √	0 – 1 1 √		
11	$ABCD$	1 1 1 1 √	– 0 1 1 √	*Note:* the same combination is NOT written down twice.	
			1 – 0 1 √		
			1 0 – 1 √		
			1 1 – 1 √		
			1 1 1 – √		
			– 1 1 1 √		
			1 – 1 1 √		

Similarly, other matched terms are 1 and 6, 2 and 3, 2 and 4, 3 and 7, 3 and 8, 4 and 7, 4 and 9, 5 and 10, 6 and 7, 6 and 10, 7 and 11, 8 and 11, 9 and 11, and, finally, 10 and 11.

The matching process is now repeated with the entries in column C and so on as shown by Table 1.8.

Two of the column D terms cannot be matched and are therefore *prime implicants*. The terms in column D that can be matched all lead to the same result, i.e. ---1.

Hence the prime implicants of the function are

$$D + B\bar{C} + AB$$

This may be the simplest form possible but a check can be carried out by drawing up a table of the prime implicants. The first step in obtaining the prime implicant table is to write down all the possible combinations for each term. Therefore

$A\ B\ -\ -$		$-\ B\ \bar{C}\ -$		$-\ -\ -\ D$	
1 1 0 0	12	0 1 0 0	4	0 0 0 1	1
1 1 0 1	13	0 1 0 1	5	0 0 1 1	3
1 1 1 0	14	1 1 0 0	12	0 1 0 1	5
1 1 1 1	15	1 1 0 1	13	0 1 1 1	7
				1 0 0 1	9
				1 0 1 1	11
				1 1 0 1	13
				1 1 1 1	15

Hence the prime implicant table is given in Table 1.9.

Table 1.9 Prime implicant table

	1	2	3	4	5	6	7	8	9	10	11	12	13	14	15
AB												√	√	√	√
$B\bar{C}$				√	√							√	√		
D	√		√		√		√		√		√		√		√

It is clear from this table that all three terms must be retained to cover all the required decimal numbers and this means that no further simplification is possible. Hence

$$F = AB + B\bar{C} + D \quad (Ans)$$

The accuracy of this result can easily be checked by means of a Karnaugh map since there are only 4 literals.

The Use of NAND/NOR Gates to Generate AND/OR Functions

The majority of the integrated circuit gates used in modern equipment belong to either the ttl or the cmos logic families. In both of these families NAND and NOR gates are cheaper and faster than the other gates which are available and, in the case of ttl, dissipate less power. It is therefore common practice to construct combinational logic circuits using either NAND or NOR gates *only*.

It is easy to see that the AND function can be obtained by cascading two NAND gates as shown by Fig. 1.6(*a*), and the OR function can be produced by the cascade connection of two NOR gates as in Fig. 1.6(*b*).

Implementation of the AND function using NOR gates, and of the OR function using NAND gates, is not quite as easy to see but the necessary connections are easily deduced with the aid of De Morgan's rules.

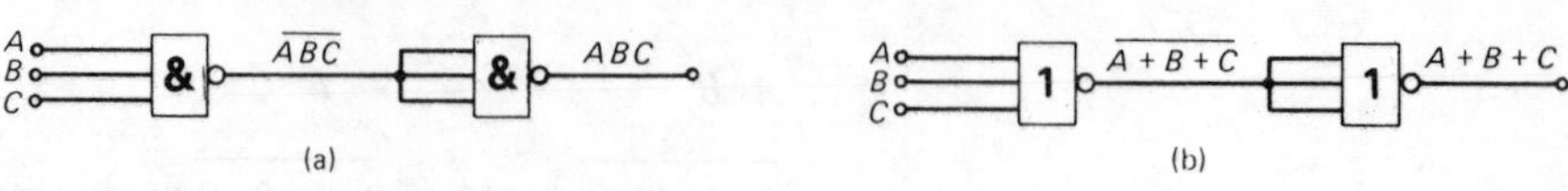

Fig. 1.6 Implementation of (*a*) the AND function using NAND gates, (*b*) the OR function using NOR gates

Rule 22 is $\overline{AB} = \bar{A} + \bar{B}$ and hence

$$AB = \overline{\bar{A} + \bar{B}}$$

The right-hand side of this equation is easily implemented using NOR gates as shown in Fig. 1.7(*a*).

The other De Morgan rule (no. 21) is $\overline{A + B} = \bar{A}\bar{B}$ and hence

$$A + B = \overline{\bar{A}\bar{B}}$$

and this can be implemented by NAND gates connected as shown by Fig. 1.7(*b*).

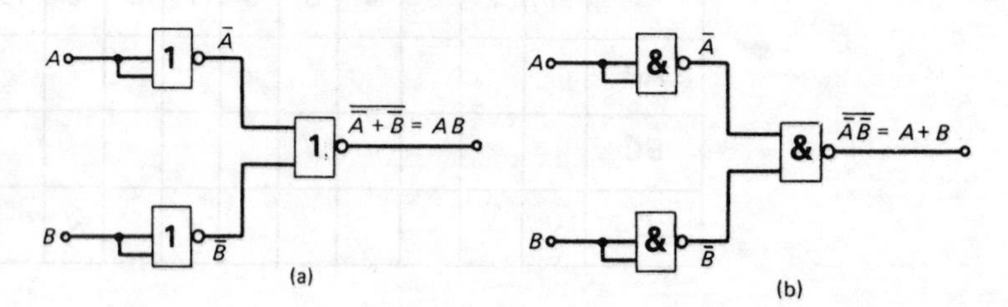

Fig. 1.7 Implementation of (*a*) the AND function using NOR gates, (*b*) the OR function using NAND gates

This principle can be extended to three, four, or more literals. For example, the function

$$F = ABCD = \overline{\bar{A} + \bar{B} + \bar{C} + \bar{D}}$$

can be fabricated using the arrangement of Fig. 1.8.

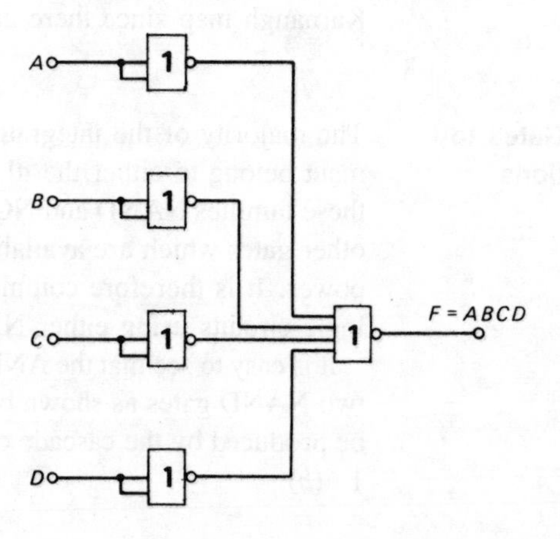

Fig. 1.8 4-input AND function implemented using NOR gates

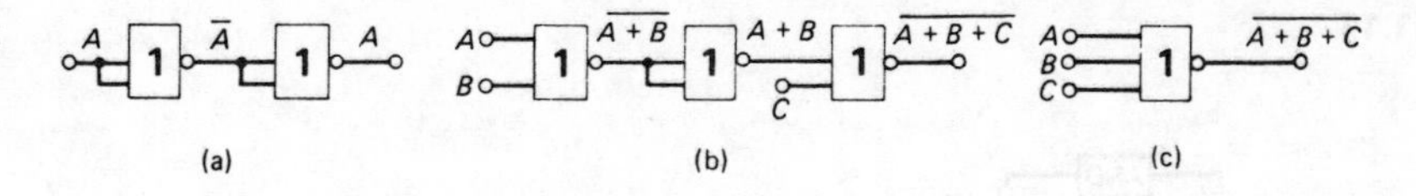

Fig. 1.9(*a*) **Redundant NOR gates, (*b*) and (*c*) two ways of implementing**

$F = \overline{A+B+C}$

Clearly, more gates are needed to implement the AND/OR functions using NAND/NOR gates but very often the apparent increase in the number of gates required is not as great as may be anticipated. This is because consecutive stages of inversion are redundant (since $\bar{\bar{A}} = A$) and need not be provided (Fig. 1.9(*a*)). Also, a circuit of the form given in Fig. 1.9(*b*) can be replaced by one 3-input gate as shown by Fig. 1.9(*c*).

In general, Boolean equations in the *product-of-sums* form are best implemented using NOR gates, and *sum-of-products* equations are more easily implemented using NAND gates. This is illustrated by the following example.

Example 1.18

Implement the logical function $F = ABC + C\bar{D}$ using *a*) NAND gates only, *b*) NOR gates only.

Solution

First draw the logic diagram using AND, OR and NOT gates and then replace each gate with its *a*) NAND or *b*) NOR equivalent circuit. Finally, simplify the circuit if possible by eliminating any redundant gates.

Fig. 1.10

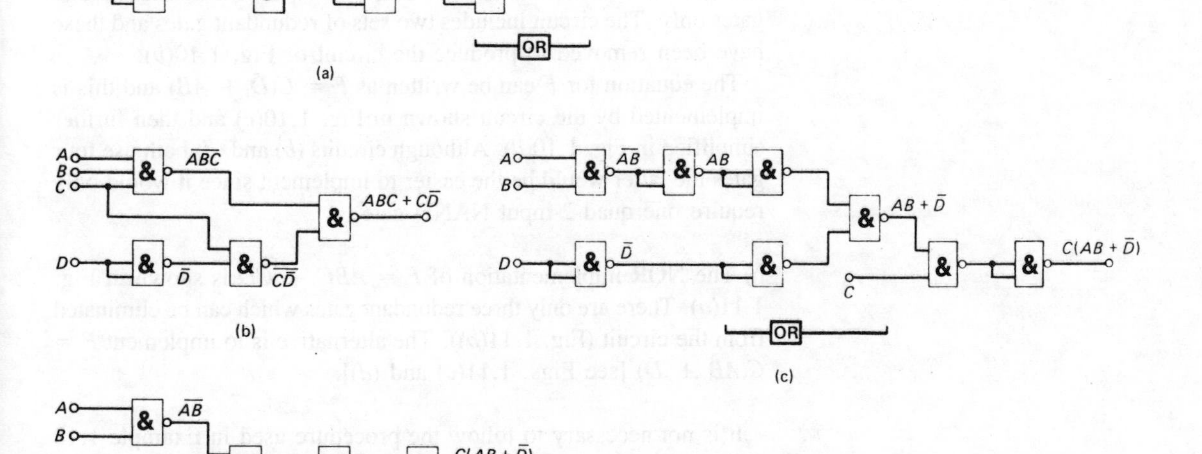

Fig. 1.11

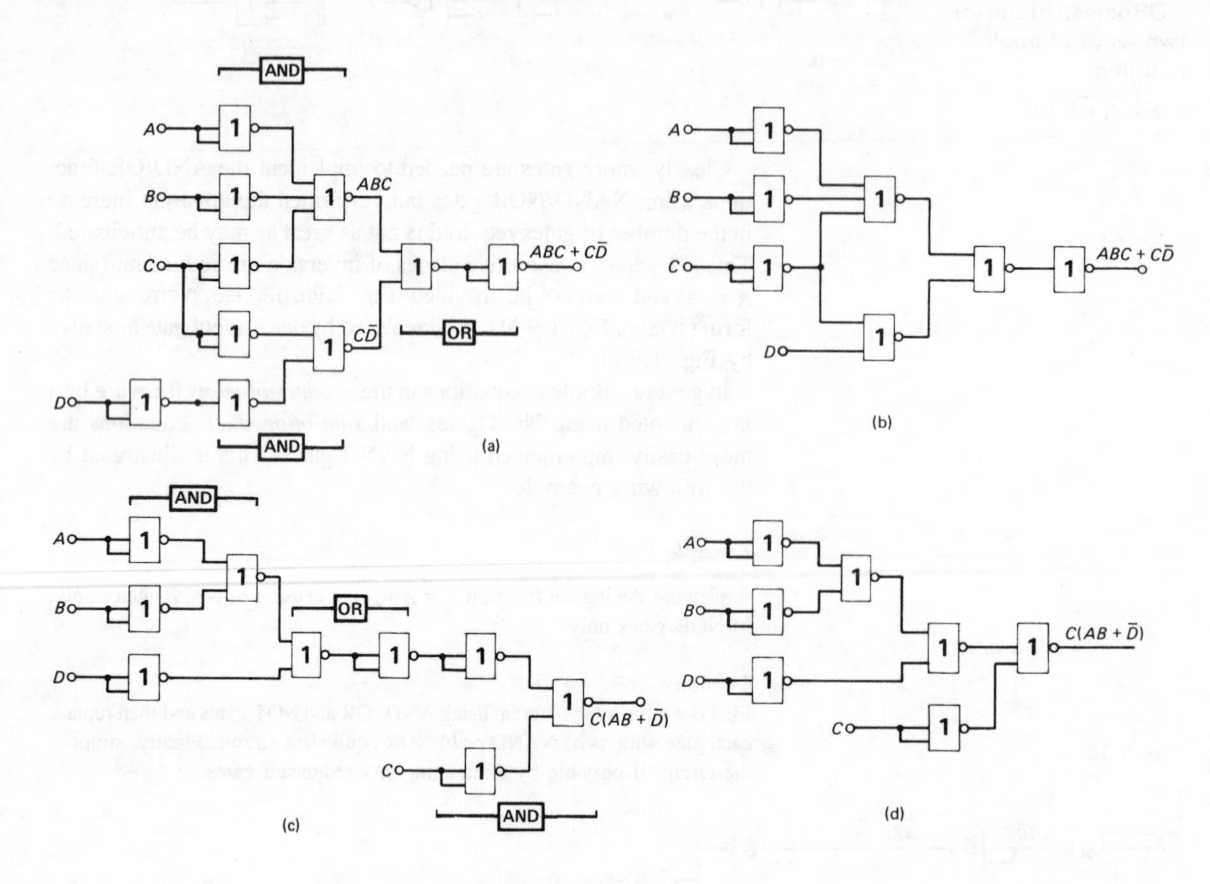

a) Figure 1.10(*a*) shows the given functions implemented using NAND gates only. The circuit includes two sets of redundant gates and these have been removed to produce the circuit of Fig. 1.10(*b*).

The equation for F can be written as $F = C(\bar{D} + AB)$ and this is implemented by the circuit shown in Fig. 1.10(*c*) and then further simplified in Fig. 1.10(*d*). Although circuits (*b*) and (*d*) both use four gates the latter would be the easier to implement since it would only require one quad 2-input NAND gate.

b) The NOR implementation of $F = ABC + C\bar{D}$ is shown in Fig. 1.11(*a*). There are only three redundant gates which can be eliminated from the circuit (Fig. 1.11(*b*)). The alternative is to implement $F = C(AB + \bar{D})$ [see Figs. 1.11(*c*) and (*d*)].

It is not necessary to follow the procedure used in Example 1.18 since some useful rules are available that allow a NAND or NOR implementation of a function to be written down directly. These rules are:

A *To implement a sum-of-products equation using NAND gates only*

a) Take the final gate(s) at the output as OR.

b) Take the *even* levels of gate, numbered from the output, as AND.

c) Take the *odd* levels of gate, numbered from the output, as OR.

d) Any literals entering the circuit at an *odd* gate level, numbered from the output, must be inverted.

The rules are illustrated by Fig. 1.12.

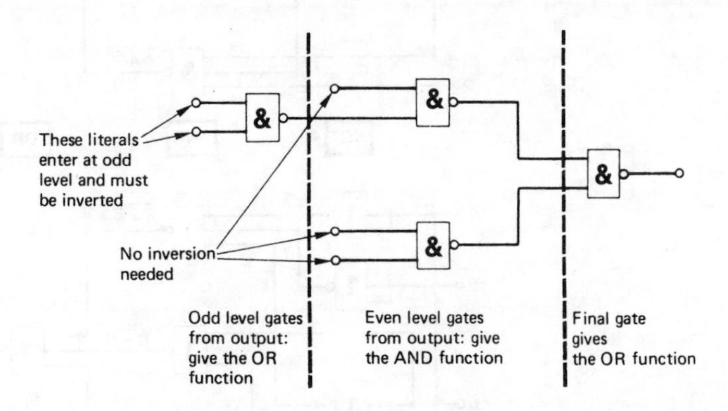

Fig. 1.12 Rules for the implementation of a sum-of-products logic function using NAND gates only

B *To implement a product-of-sums equation using NOR gates only*

a) Take the final gate at the output as AND.

b) Take the *even* levels of gate, numbered from the output, as OR.

c) Take the *odd* levels of gate, numbered from the output, as AND.

d) Any literals entering at an *odd* level must be inverted.

Example 1.19

Implement using *a*) NAND gates only, *b*) NOR gates only, the function

$$F = \bar{A}C + \bar{B}C + A\bar{C}D$$

Solution

a) Using rule **A**, Fig. 1.13(*a*) can be drawn directly. To check:

$$F = \overline{(\overline{\bar{A}C})(\overline{\bar{B}C})(\overline{A\bar{C}D})} = \bar{A}C + \bar{B}C + A\bar{C}D$$

b) The mapping of the function F is

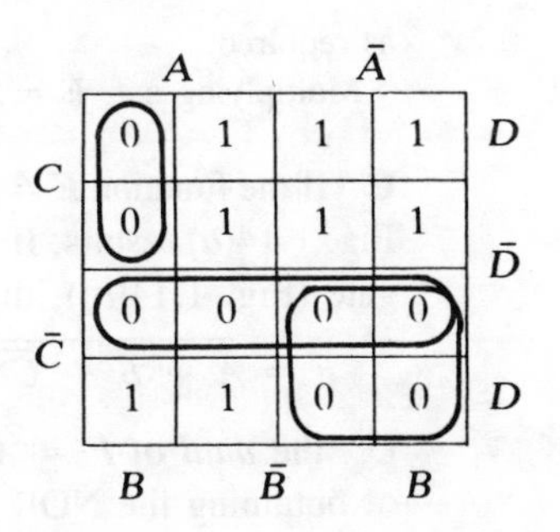

Fig. 1.13

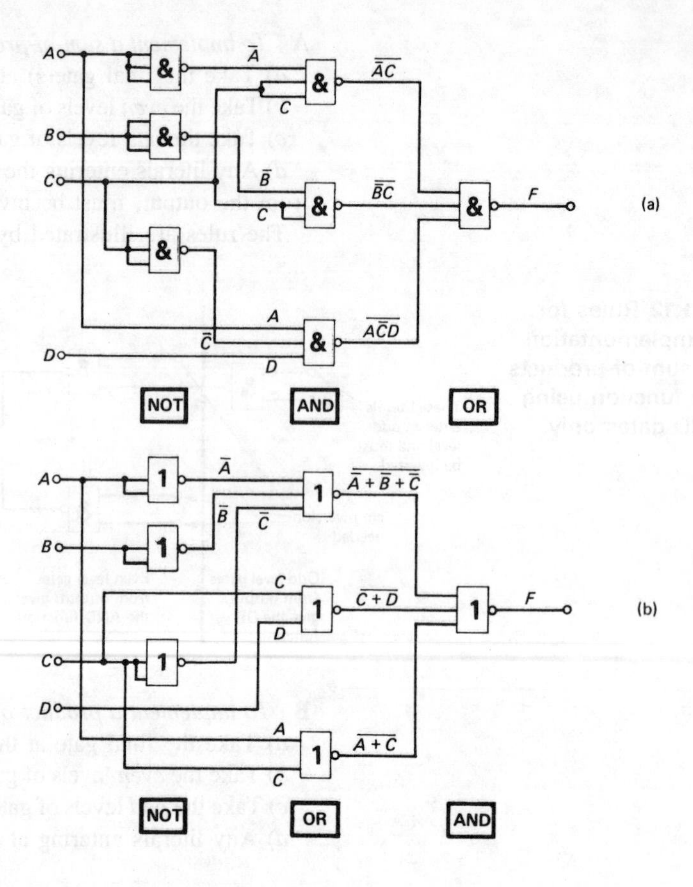

From the looped 0 squares $\bar{F} = ABC + \bar{C}\bar{D} + \bar{A}\bar{C}$

Hence $F = \overline{ABC + \bar{C}\bar{D} + \bar{A}\bar{C}}$

$$= (\overline{ABC})(\overline{\bar{C}\bar{D}})(\overline{\bar{A}\bar{C}})$$

$$= (\bar{A} + \bar{B} + \bar{C})(C + D)(A + C)$$

Using rule **B**, Fig. 1.13(*b*) can be drawn. Checking

$$F = \overline{\overline{(\bar{A} + \bar{B} + \bar{C})} + \overline{(C + D)} + \overline{(A + C)}}$$

$$= (\bar{A} + \bar{B} + \bar{C})(C + D)(A + C)$$

as required.

Multiplying out, $F = \bar{A}C + \bar{B}C + A\bar{C}D$, the original equation.

C If the function $F = AB + CD$ is implemented using NAND gates, Fig. 1.14(*a*) results. If, now, each NAND gate is replaced by a NOR gate (Fig. 1.14(*b*)), the output F from the circuit is

$$F = \overline{\overline{A + B} + \overline{C + D}} = (A + B)(C + D)$$

i.e. the *dual* of $F = AB + CD$. This result provides another way of obtaining the NOR gate implementation of a Boolean equation.

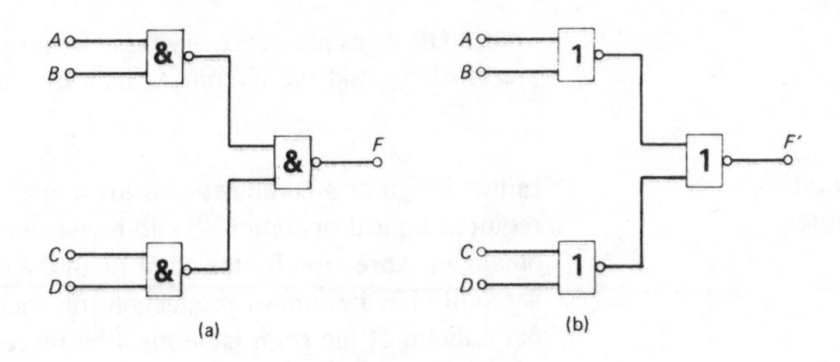

Fig. 1.14 Replacing NAND gates with NOR gates gives the dual of a function

Consider once again the function given in Example 1.19. The dual of this equation is

$$F' = (\bar{A} + C)(\bar{B} + C)(A + \bar{C} + D)$$

Multiplying out, $F' = AC + \bar{A}\bar{B}\bar{C} + CD$

This equation can be implemented by NAND gates (Fig. 1.15(*a*)) and then each gate can be replaced by a NOR gate (Fig. 1.15(*b*)). The output F of the final NOR gate is

$$\begin{aligned} F &= \overline{\overline{A + C} + \overline{C + D} + \overline{\bar{A} + \bar{B} + \bar{C}}} \\ &= (A + C)(C + D)(\bar{A} + \bar{B} + \bar{C}) \\ &= \bar{A}C + \bar{B}C + A\bar{C}D \quad \text{as before} \end{aligned}$$

Figs. 1.13(*b*) and 1.15(*b*) should be compared.

D Another method consists of *doubly* inverting the equation to be implemented. Thus, for the equation $F = \bar{A}C + \bar{B}C + A\bar{C}D$ (again),

$$\bar{F} = \overline{\bar{A}C + \bar{B}C + A\bar{C}D} = (\overline{\bar{A}C})(\overline{\bar{B}C})(\overline{A\bar{C}D}) \qquad \text{and}$$

$$F = \overline{(\overline{\bar{A}C})(\overline{\bar{B}C})(\overline{A\bar{C}D})}.$$

This equation shows immediately that, if the complements of literals A, B and C are available, four NAND gates are required. The implementation is shown by Fig. 1.13(*a*).

Similarly, if $F = (\bar{A} + \bar{B} + \bar{C})(C + D)(A + C)$

$$\bar{F} = (\overline{\bar{A} + \bar{B} + \bar{C}}) + (\overline{C + D}) + (\overline{A + C})$$

$$F = \overline{(\overline{\bar{A} + \bar{B} + \bar{C}}) + (\overline{C + D}) + (\overline{A + C})}.$$

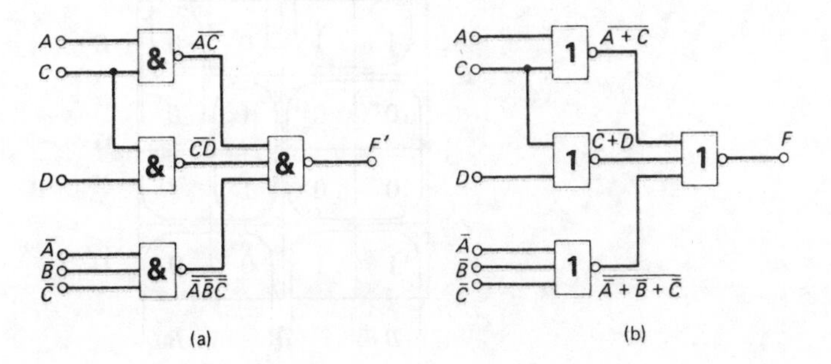

Fig. 1.15 NOR implementation of $F = \bar{A}C + \bar{B}C + A\bar{C}D$

Four NOR gates are necessary (assuming the required complements are available and the circuit is given by Fig. 1.13(*b*)).

Designing a Circuit from a Truth Table

In the design of a combinational logic circuit, the truth table of the required logical operation should be written down and then used to obtain an expression for the output F of the circuit. If the expression is required in the sum-of-products form, each 1 appearing in the output column of the truth table must be represented by a term in the Boolean equation describing the circuit. Every such term must contain each literal that is in the logical 1 state and the complement of each literal that is the logical 0 state. This expression can then be simplified, using one of the methods presented earlier, before it is implemented by the suitable interconnection of a number of gates.

Example 1.20

Design a circuit that will indicate whether a 4-bit number is either odd and greater than 8 or even and less than 5. Assume decimal 0 to be an even number. Implement the circuit using either NAND or NOR gates only.

Solution

Table 1.10 gives the truth table for the wanted circuit.

Table 1.10

A	0	1	0	1	0	1	0	1	0	1	0	1	0	1	0	1
B	0	0	1	1	0	0	1	1	0	0	1	1	0	0	1	1
C	0	0	0	0	1	1	1	1	0	0	0	0	1	1	1	1
D	0	0	0	0	0	0	0	0	1	1	1	1	1	1	1	1
F	1	0	1	0	1	0	0	0	0	1	0	1	0	1	0	1

From the table,

$$F = \bar{A}\bar{B}\bar{C}\bar{D} + \bar{A}B\bar{C}\bar{D} + \bar{A}\bar{B}C\bar{D} + A\bar{B}\bar{C}D + AB\bar{C}D + A\bar{B}CD + ABCD$$

Mapping:

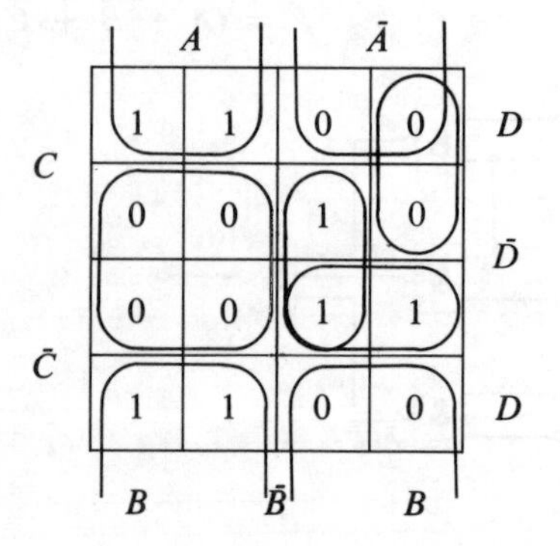

From the map

$$F = AD + \bar{A}\bar{B}\bar{D} + \bar{A}\bar{C}\bar{D} \text{ or } F = (\bar{A} + D)(A + \bar{D})(A + \bar{B} + \bar{C})$$

The required circuits are shown in Fig. 1.16(*a*) and (*b*).

Fig. 1.16

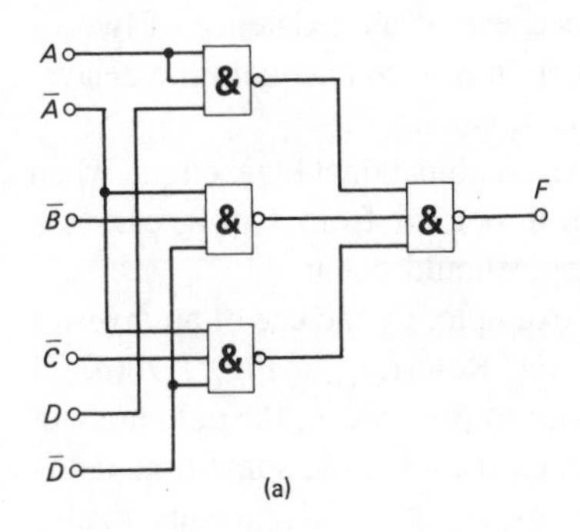

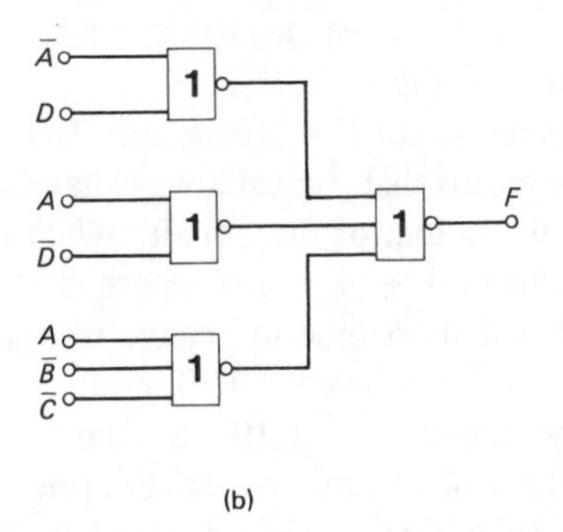

Example 1.21

A circuit is required that has ten inputs, numbered 0 through to 9, and one output. The output is to be high whenever any one or more of the inputs numbered 1, 3, 7 or 9 are high. Design the circuit using *a*) NAND gates and *b*) NOR gates only.

Solution

The truth table for the circuit is shown by Table 1.11.

From the table

$$F = A\bar{B}\bar{C}\bar{D} + AB\bar{C}\bar{D} + ABC\bar{D} + A\bar{B}\bar{C}D$$

and its mapping is

	A		$\bar{A}$		
	B	$\bar{B}$	$\bar{B}$	B	
C	X	X	X	X	D
C	1	0	0	0	$\bar{D}$
$\bar{C}$	1	1	0	0	$\bar{D}$
$\bar{C}$	X	1	0	X	D

Table 1.11

Input	0	1	2	3	4	5	6	7	8	9
A	0	1	0	1	0	1	0	1	0	1
B	0	0	1	1	0	0	1	1	0	0
C	0	0	0	0	1	1	1	1	0	0
D	0	0	0	0	0	0	0	0	1	1
Output *F*	0	1	0	1	0	0	0	1	0	1

Fig. 1.17

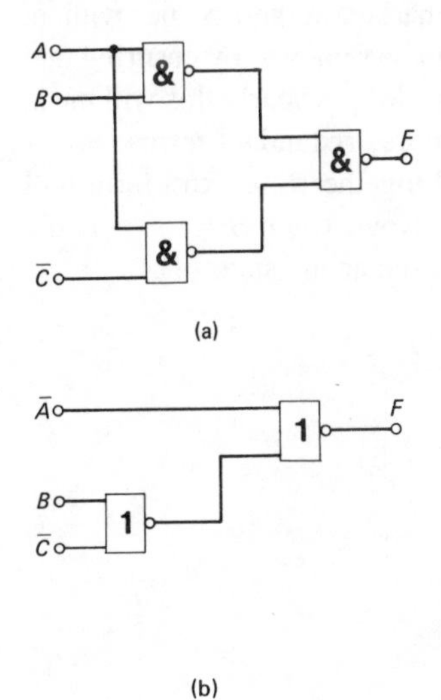

a) From the map (1s and Xs), $F = AB + A\bar{C}$ and this is implemented in Fig. 1.17(*a*).

b) From the map (0s and Xs), $F = A(B + \bar{C})$ and this is implemented by the circuit of Fig. 1.17(*b*).

Race Hazards

A **race hazard** in a combinational logic circuit is an unwanted transient that produces a spike or *glitch* at the output of the circuit. To

Fig. 1.18 (*a*) Circuit producing a race hazard, (*b*) waveforms for circuit in (*a*)

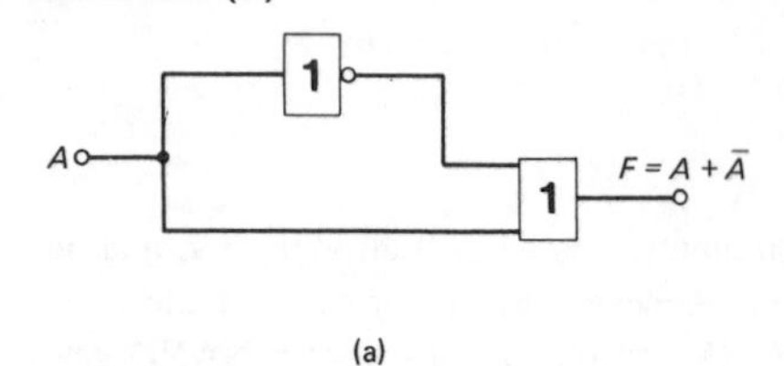

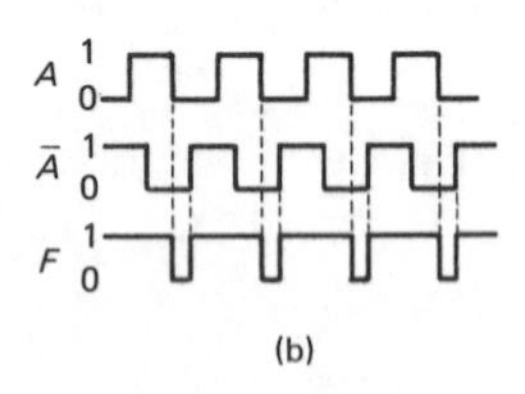

Fig. 1.19 Circuit producing $F = \bar{A}B + AC$ gives unwanted output when A changes from 1 to 0

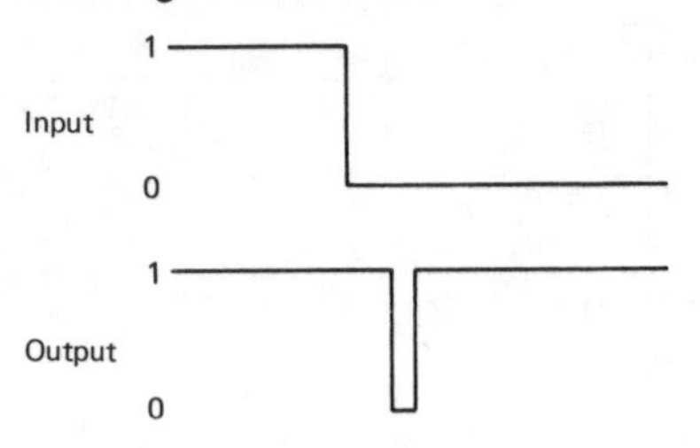

put it alternatively: a hazard is said to exist if a single input change can result in the occurrence of one or more unwanted momentary outputs. The hazard usually results because of the existence of two or more paths through the circuit which introduce unequal time delays. The hazard may be either *static* or *dynamic*.

A **static hazard** is said to exist in a combinational logic circuit when the change of a single literal from 0 to 1, or from 1 to 0, causes a glitch at the output when no change should occur.

A hazard may be produced, for example, by the use of an inverter to obtain the complement of a literal. Referring to Fig. 1.18(*a*); if the literal A is applied to the inverter to produce $\bar{A}$, the action is not instantaneous. Hence, if A is changed from 0 to 1, some time delay is inevitably introduced in A becoming available. This means that for some short time $A + \bar{A}$ is *not* equal to 1 and consequently a false output is generated for a short time (see Fig. 1.18(*b*).

Consider a circuit which produces an output $F = \bar{A}B + AC$. The use of an inverting stage to obtain $\bar{A}$ means that $\bar{A}$ must always have a small time delay relative to A at the output of the circuit. When $B = C = 1$, then F should be of the form $\bar{A} + A = 1$. Suppose that A changes from 1 to 0. Then $\bar{A}$ takes a short time to change from 0 to 1, and hence for a short while $A + \bar{A}$ will be equal to 0, producing a short, unwanted 0 spike at the output (Fig. 1.19). Similarly, when A changes from 0 to 1, a short period of time will exist during which *both* A and $\bar{A}$ will be a logical 1 but this will not affect the output.

A static hazard exists in a circuit if the literals can alter to produce a change between adjacent cells in the Karnaugh mapping of the functions that are *not* looped together. Thus, in the mapping for $\bar{A}B + AC$, a hazard exists since the squares marked in their corner with an X are not looped together. A hazard can be removed by ensuring that such adjacent cells *are* included within a loop. Clearly this will mean the introduction of one, or perhaps more, redundant terms.

If the squares marked X are looped together, the extra term will be BC, so that $F = \bar{A}B + AC + BC$. Now, when $B = C = 1$, the output will be 1 regardless of the instantaneous state of A.

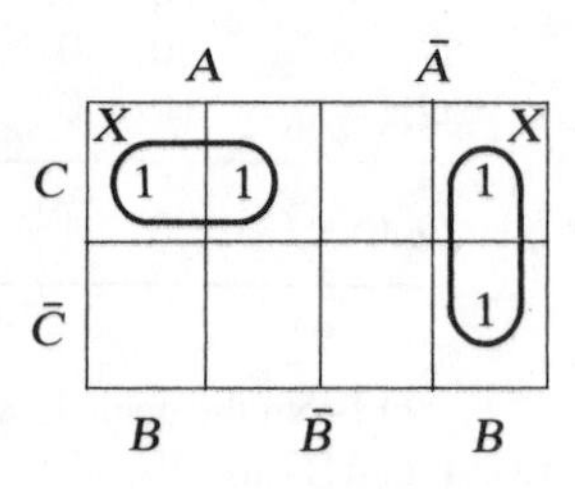

Example 1.22

The mapping of a Boolean equation is shown opposite. Obtain *a*) the minimal solution, *b*) the race-free solution of the equation.

	A	A	$\bar{A}$	$\bar{A}$	
C	0	1	1	0	D
C	0	1	1	0	$\bar{D}$
$\bar{C}$	1	1	0	0	$\bar{D}$
$\bar{C}$	0	0	1	1	D
	B	$\bar{B}$	$\bar{B}$	B	

Solution
The minimal solution is shown by map A and is

$$F = \bar{B}C + A\bar{C}\bar{D} + \bar{A}\bar{C}D \quad (Ans)$$

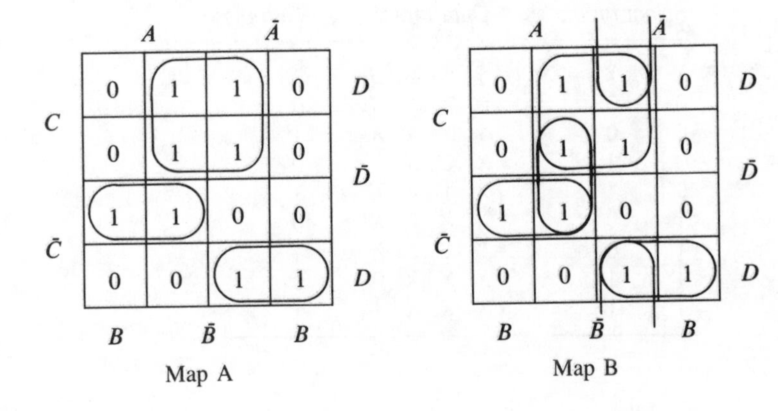
Map A Map B

The race-free solution is obtained from map B and is

$$F = \bar{B}C + A\bar{C}\bar{D} + \bar{A}\bar{C}D + \bar{A}\bar{B}D + A\bar{B}\bar{D} \quad (Ans)$$

A **dynamic hazard** is said to exist when the output of a circuit changes two or more times when it should have changed once only in response to an input change. An example of a dynamic hazard is shown by Fig. 1.20.

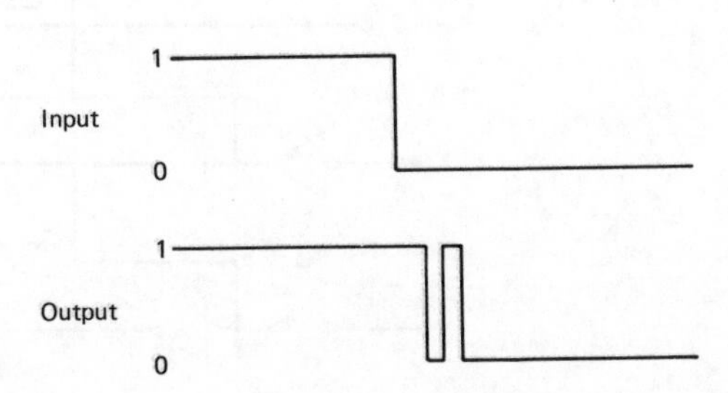

Fig. 1.20 Illustrating a dynamic hazard

Logic Design Using Multiplexers

The availability of integrated circuit devices known as multiplexers provides an alternative to the use of gates for the implementation of logic functions. The equation to be implemented must first be expressed in its canonical sum-of-products form.

The **multiplexer** or **data selector** is a circuit which will select *one* out of *n* input lines where *n* is a power of 2. The design of such circuits using separate logic gates and inverters is considered in Chapter 3 (p. 93). Usually an msi device would be used, several of which are available in both the ttl and cmos families. The block diagram of one-half of a dual 4-line to 1-line multiplexer, the ttl 74LS153, is shown by Fig. 1.21. When the enable input is low, the circuit is able to switch any of four inputs S_0, S_1, S_2, S_3 to the output terminal. The truth table describing the circuit's logic operation is given by Table 1.12 where *X* indicates the don't-care condition.

Table 1.12 Multiplexer truth table (4 inputs)

Select Inputs		*Data Inputs*				*Outputs*
X	*Y*	S_0	S_1	S_2	S_3	
0	0	0	*X*	*X*	*X*	0
0	0	1	*X*	*X*	*X*	1
0	1	*X*	0	*X*	*X*	0
0	1	*X*	1	*X*	*X*	1
1	0	*X*	*X*	0	*X*	0
1	0	*X*	*X*	1	*X*	1
1	1	*X*	*X*	*X*	0	0
1	1	*X*	*X*	*X*	1	1

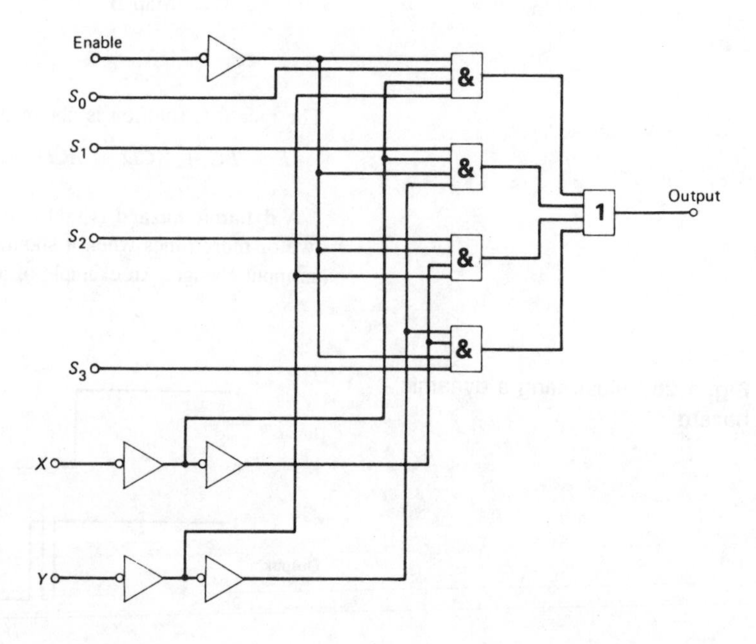

Fig. 1.21 4-to-1 multiplexer

When both the select inputs X and Y are low, the gate connected to S_0 input is the only one that is not inhibited by either X or Y and so the output takes up the S_0 state. Similarly, any other combination of the select input states enables just one of the four AND gates and connects the output terminal to the appropriate input terminal.

From Table 1.12, the Boolean equation describing the operation of the circuit is

$$F = \bar{X}\bar{Y}S_0 + \bar{X}YS_1 + X\bar{Y}S_2 + XYS_3 \qquad (1.6)$$

There are four possible values that can be applied to each of the four data input lines: these are 1, 0, C and $\bar{C}$ where C is a literal. There are two literals, A and B, which can be applied to the select input lines. With three literals A, B and C, there are three different ways in which the select variables can be chosen, i.e. AB, AC or BC.

For each possibility a Karnaugh map can be produced which will show the various positions of the data inputs S_0, S_1, S_2, S_3.

Suppose, firstly, that A and B are chosen to be the select input or *control* variables, then equation (1.6) becomes

$$F = \bar{A}\bar{B}S_0 + \bar{A}BS_1 + A\bar{B}S_2 + ABS_3$$

and the Karnaugh mapping is as shown by Map A.

Map A

	A		$\bar{A}$	
C	S_3	S_2	S_0	S_1
$\bar{C}$	S_3	S_2	S_0	S_1
	B	$\bar{B}$		B

Similarly, if A and C are chosen to be the control variables, equation (1.6) is written as

$$F = \bar{A}\bar{C}S_0 + \bar{A}CS_1 + A\bar{C}S_2 + ACS_3$$

and this is mapped by Map B. Lastly, Map C gives the mapping obtained if B and C are the control variables.

Map B

	A		$\bar{A}$	
C	S_3	S_3	S_1	S_1
$\bar{C}$	S_2	S_2	S_0	S_0
	B	$\bar{B}$		B

Map C

	A		$\bar{A}$	
C	S_3	S_2	S_2	S_3
$\bar{C}$	S_1	S_0	S_0	S_1
	B	$\bar{B}$		B

In the design of a multiplexer solution to a logic problem, an arbitrary choice must first be made as to which two of the three literals A, B, and C are to be chosen as the control variables and then the appropriate Map (A, B or C) will be used.

Suppose the Boolean equation

$$F = ABC + \bar{A}\bar{B}C + \bar{A}B\bar{C} + A\bar{B}\bar{C}$$

is to be implemented using a 4-input multiplexer. The mapping of this equation is given by Map D.

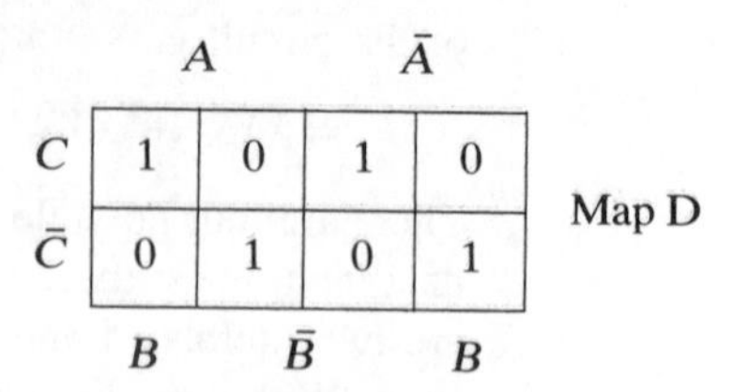

	A	A	Ā	Ā
C	1	0	1	0
C̄	0	1	0	1
	B	B̄	B̄	B

Map D

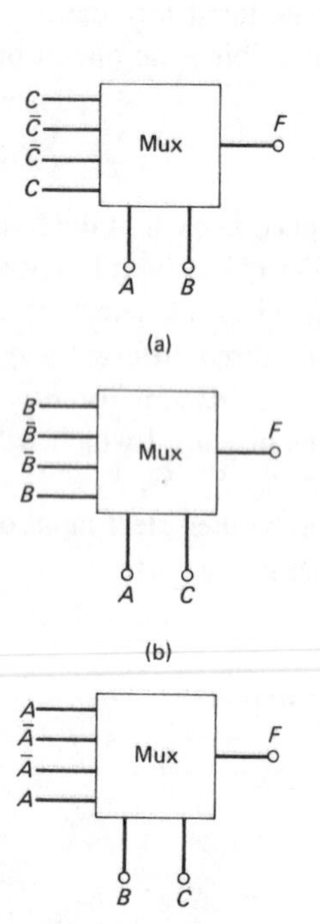

Fig. 1.22 Multiplexer solutions to $F = ABC + \bar{A}\bar{B}C + \bar{A}B\bar{C} + A\bar{B}\bar{C}$

If A and B are chosen to be the control variables this map must be compared with Map A. Hence,

$$S_0 = C + 0 = C = S_3 \qquad S_1 = S_2 = \bar{C}$$

and the required circuit is shown by Fig. 1.22(a).

If the control variables are chosen to be A and C, Maps B and D must be compared. Now,

$$S_0 = B = S_3 \qquad S_1 = S_2 = \bar{B} \quad \text{(see Fig. 1.22(}b\text{))}$$

Lastly, for control variables B and C,

$$S_0 = A \qquad S_1 = \bar{A} \qquad S_2 = \bar{A} \qquad S_3 = A \qquad \text{(Fig. 1.22(}c\text{))}$$

The procedure can be extended to functions with four literals. Suppose the function

$$F = ABCD + \bar{A}\bar{B}CD + A\bar{B}C\bar{D} + AB\bar{C}\bar{D} + A\bar{B}\bar{C}D$$

is to be implemented. As before, two of the literals must be arbitrarily chosen to be the control variables: here A and B have been so selected. The 16-square Karnaugh map of the basic multiplexer equation (1.6) is given by Map E (an extension of Map A), while Map F represents the equation to be implemented. Comparing the two tables gives

$$S_0 = CD \qquad S_1 = 0 \qquad S_2 = C\bar{D} + \bar{C}D \qquad S_3 = CD + \bar{C}\bar{D}$$

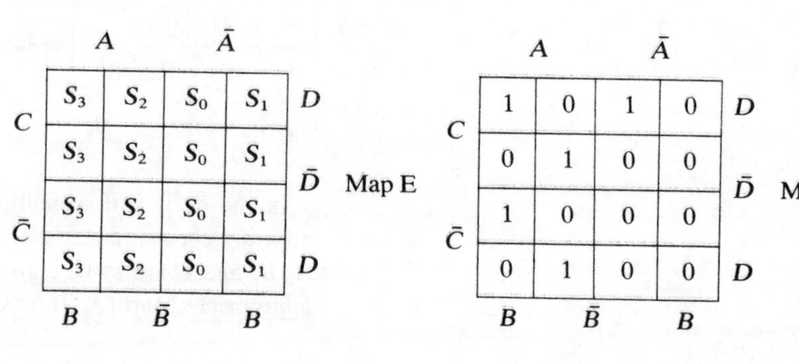

	A	A	Ā	Ā	
C	S_3	S_2	S_0	S_1	D
C	S_3	S_2	S_0	S_1	D̄
C̄	S_3	S_2	S_0	S_1	D̄
C̄	S_3	S_2	S_0	S_1	D
	B	B̄	B̄	B	

Map E

	A	A	Ā	Ā	
C	1	0	1	0	D
C	0	1	0	0	D̄
C̄	1	0	0	0	D̄
C̄	0	1	0	0	D
	B	B̄	B̄	B	

Map F

Alternatively, the equation can be implemented using an 8-input multiplexer such as the 74LS151. The truth table of such a device is given by Table 1.13.

Table 1.13 Multiplexer truth table (8 inputs)

Select inputs			*Data inputs*								*Output*
X	Y	Z	S_0	S_1	S_2	S_3	S_4	S_5	S_6	S_7	
0	0	0									S_0
1	0	0									S_1
0	1	0									S_2
1	1	0									S_3
0	0	1									S_4
1	0	1									S_5
0	1	1									S_6
1	1	1									S_7

Only the output has been given for each combination of select inputs to simplify the table. The Boolean equation describing this operation is

$$F = S_0\bar{X}\bar{Y}\bar{Z} + S_1X\bar{Y}\bar{Z} + S_2\bar{X}Y\bar{Z} + S_3XY\bar{Z} + S_4\bar{X}\bar{Y}Z + S_5X\bar{Y}Z + S_6\bar{X}YZ + S_7XYZ \qquad (1.7)$$

Comparing equation (1.7) with the equation to be implemented it can be seen that, letting $A = X$, $B = Y$, $C = Z$, there is no term in $\bar{A}\bar{B}\bar{C}$ and so $S_0 = 0$. Similarly, the fifth term of the function, i.e. $A\bar{B}\bar{C}D$, corresponds with the second term of (1.7), hence $S_1 = D$. Carrying on this way, $S_2 = 0$, $S_3 = \bar{D}$, $S_4 = D$, $S_5 = \bar{D}$, $S_6 = 0$, and $S_7 = D$.

Example 1.23

Implement, using an 8-input multiplexer, the Boolean equation

$$F = \bar{A} + B\bar{C} + \bar{B}C$$

Solution

Writing the equation in its canonical form:

$$F = \bar{A}(B + \bar{B})(C + \bar{C}) + (A + \bar{A})B\bar{C} + (A + \bar{A})\bar{B}C$$

$$= \bar{A}BC + \bar{A}\bar{B}C + \bar{A}B\bar{C} + \bar{A}\bar{B}\bar{C} + AB\bar{C} + A\bar{B}C$$

Comparing with Table 1.13:

$$S_0 = 1,\ S_1 = 0,\ S_2 = 1,\ S_3 = 1,\ S_4 = 1,\ S_5 = 1,\ S_6 = 1,\ S_7 = 0$$

Wired-OR and AND-OR-Invert Logic

Some of the gates in the ttl logic family are designed as *open-collector* types. These are gates that can have their outputs directly paralleled and connected via a *pull-up resistor* to the positive supply voltage (see Fig. 1.23). The outputs of the individual gates are $\overline{AB}$ and $\overline{CD}$; if the output of either gate goes to zero volts the output of the paralleled

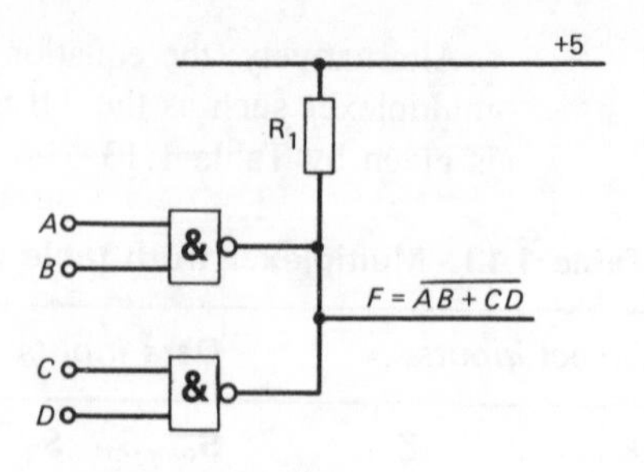

Fig. 1.23 Direct connection of open-collector gates

gates must also become 0 V. Only if both outputs are at logical 1 can the combined output be 1. The logical function performed by a **wired-OR gate** is hence

$$\overline{AB}\,\overline{CD} = \overline{AB + CD}$$

Suppose the function $F = AC + BC$ is to be implemented using wired-OR logic. The mapping of this function is

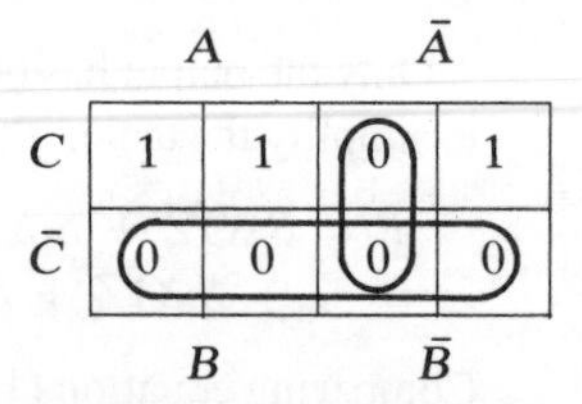

The inputs to the wired-OR gates should be the terms that correspond to the looped 0s in the mapping, i.e. $\bar{C}$ and $\bar{A}\bar{B}$, as shown by Fig. 1.24.

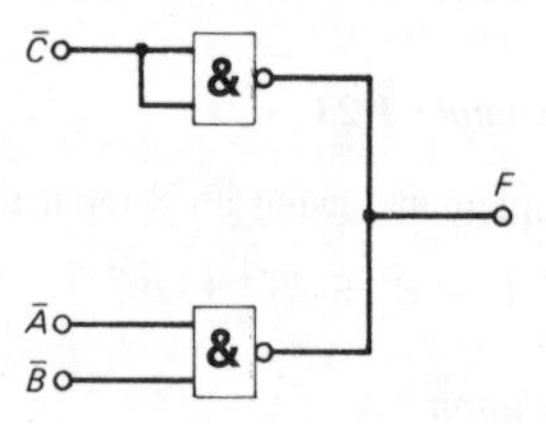

Fig. 1.24 Implementation of $F = AC + BC$ using wired-OR logic

NOR gates do not give a similar result; if open-circuit collector NOR gates have their output terminals paralleled, the effect is to increase their *fan-out* capability.

Gates with totem-pole output circuits (p. 49) must not be connected together in this way, since if only one gate is turned ON the resulting current may well damage one of the gates.

The **AND-OR-Invert (AOI) gate**, available in both the ttl and cmos logic families, provides the logic circuit shown in Fig. 1.25(*a*). The logical function performed by the AOI gate is

$$F = \overline{AB + CD}$$

The term *wide* indicates the number of inputs to the NOR gate.

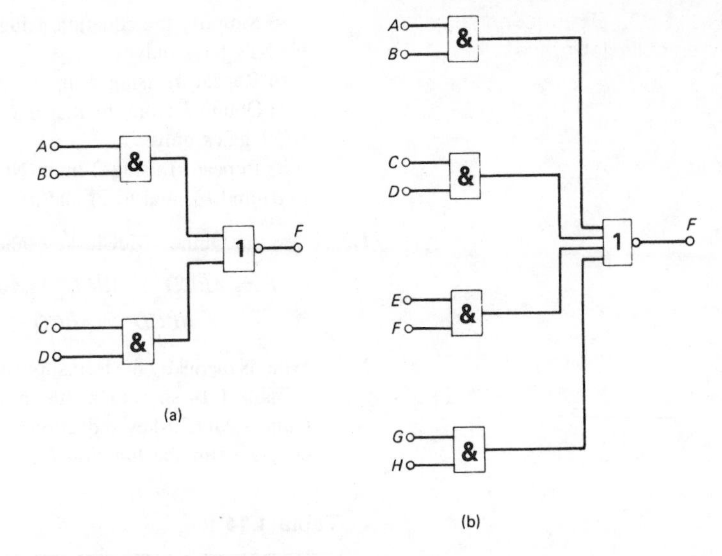

Fig. 1.25 (*a*) 2-wide AOI gate, (*b*) 4-wide 2-input AOI gate

Note that this is the same as the function provided by the wired-OR circuit. The circuit is said to be a 2-wide 2-input AOI gate because there are two 2-input AND gates working to the output NOR gate. Similarly, a 4-wide 2-input AOI gate would have four 2-input AND gates (Fig. 1.25(*b*)). However, some ic gates may have more inputs than their name indicates; for example the ttl dual 2-wide 2-input AOI gate 74LS51 has one gate as per Fig. 1.25(*a*) but the other gate includes two 3-input AND gates.

Exercises 1

1.1 What is meant by duality in Boolean algebra? Find the complements of the equations

(i) $F = C(AB + \bar{A}\bar{B})$
(ii) $F = AC(A + B)(\bar{A} + \bar{B})$

by *a)* the use of duality, *b)* using a Karnaugh map. Implement the complements using either NAND or NOR gates only.

1.2 What are race hazards in a logic circuit? Plot the function

$$F = ABCD + \bar{A}\bar{C}\bar{D} + A\bar{B}D + A\bar{C}D + \bar{A}C\bar{D}$$

on a Karnaugh map. Thence obtain *a)* the minimal solution, *b)* the simplest race-free solution. Implement both *a)* and *b)* using either NAND or NOR gates only.

1.3 The function $F = C(A + \bar{A}B + \bar{A}\bar{B}D)$ is to be implemented using (i) NAND gates, (ii) NOR gates only.

Simplify the equation and draw suitable circuits. What is meant by wired-OR logic? Could it be used to implement F?

1.4 The function $F = CD + \bar{B}\bar{C} + \bar{A}BD$ is to be implemented.

a) Simplify the equation using a Karnaugh map and implement its using NAND gates only.

b) Repeat *a*) using 2-input NAND gates only.

c) Obtain $\bar{F}$ from the map and invert to get F. Implement the result using NOR gates only.

d) Repeat *c*) using 2-input NOR gates only. Comment on your answers to *a*) and *b*), and to *c*) and *d*).

1.5 Use the Quine—McCluskey tabular method to simplify the equation

$$F = \bar{A}\bar{B}CD + A\bar{B}CD + A\bar{B}\bar{C}D + \bar{A}BCD + \bar{A}B\bar{C}D + \bar{A}B\bar{C}\bar{D} + \bar{A}\bar{B}C\bar{D} + A\bar{B}C\bar{D} + A\bar{B}\bar{C}\bar{D}$$

1.6 What is meant by the terms: prime implicant, literal, minterm and maxterm?

Table 1.14 shows a prime implicant chart which has been produced by a Quine—McCluskey reduction of a Boolean equation. Obtain the minimum coverage for the function F.

Table 1.14

	0	1	2	3	4	5	6	7	8	9	10	11	12	13	14	15
$\bar{A}\bar{B}\bar{C}$	√		√											√		
AB			√								√					
BCD	√										√	√				

1.7 Implement the function $F = (\overline{A + B + C})(AB + \bar{A}\bar{B} + AC) + ABC$ using *a*) AOI gates, *b*) NAND gates only, *c*) NOR gates only.

1.8 A circuit is required, using the minimum number of NAND gates only, that has two outputs F_1 and F_2 where

$$F_1 = \bar{B}C\bar{D} + \bar{A}\bar{B}CD + \bar{A}\bar{B}\bar{C}D$$

$$F_2 = A\bar{B}C\bar{D} + \bar{A}BD + \bar{A}BC\bar{D}$$

Design the circuit.

1.9 Map the function $\Sigma(1, 3, 5, 7, 9, 11, 12)$ with don't-care conditions 2, 8 and 10. Obtain two simplified Boolean equations, one suitable for NAND gate, the other for NOR gate implementation. Draw the two logic circuits. Obtain the Boolean equations describing the circuit whose truth table is given by Table 1.15. Simplify the equation and then implement it using (i) NAND gates only, (ii) NOR gates only.

Table 1.15

A	0	0	0	0	1	1	1	1
B	0	0	1	1	0	0	1	1
C	0	1	0	1	0	1	0	1
F	1	1	1	1	0	0	0	1

1.10 Implement the function

$$F = ACD + \bar{A}BCD + \bar{B}\bar{C}D + \bar{A}\bar{B}C\bar{D} + \bar{A}\bar{B}CD + A\bar{B}C\bar{D}$$

using (i) a 4-input multiplexer, (ii) an 8-input multiplexer.

Fig. 1.26

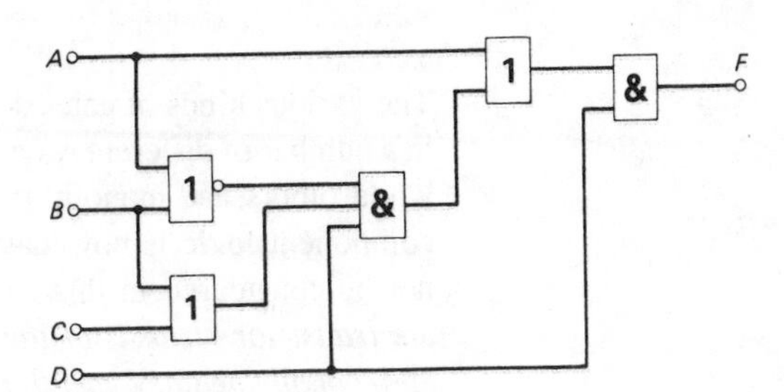

1.11 Determine the Boolean expression for the output of the circuit shown in Fig. 1.26. Simplify the expression if possible and then implement the circuit using (i) NAND gates, (ii) NOR gates only.

1.12 Implement the function given in 1.11 using an 8-input multiplexer.

1.13 Simplify, using a Karnaugh map,

a) $F = \bar{A}\bar{B} + \bar{A}B + A\bar{B}$
b) $F = AB + \bar{A}BC + \bar{B}\bar{C}$
c) $F = (\bar{A} + \bar{B})(\bar{A} + B)(A + B)$
d) $F = (AB + \bar{B}\bar{C})(\bar{A}B + C)$

1.14 Show that

a) $A + \bar{A}B = AB$
b) $AB + \bar{B}C + AB\bar{D} + A\bar{B}C + ABCE = AB + \bar{B}C$
c) $AC + \bar{B}\bar{C}D + B\bar{C}\bar{D} + \bar{A}\bar{B}D + \bar{A}\bar{B}\bar{C}\bar{D} + A\bar{B}\bar{C}\bar{D}$
$= AC + CD + \bar{C}\bar{D} + \bar{C}\bar{B}$

1.15 *a*) Express $F = \overline{\overline{A + BC} + (A + \bar{C})B}$ in sum-of-products form.
b) Express $F = AB + AC + B\bar{C}$ in product-of-sums form.

1.16 Implement the function $F = A\bar{B}C + \bar{A}B\bar{C}$ using NOR gates only.

1.17 Implement the function $F = A\bar{B}C + \bar{A}B\bar{C} + A\bar{B}\bar{C} + \bar{A}\bar{B}C$ using a 4-input multiplexer.

1.18 Plot the function $F = AC\bar{D} + \bar{B}\bar{C}\bar{D} + \bar{B}C\bar{D} + \bar{A}B\bar{C} + \bar{A}\bar{C}D$ on a Karnaugh map and indicate where static hazards exist. State how the hazards could be eliminated.

1.19 Simplify both algebraically and by mapping each of the following:

a) $F = AB + AC + BC$
b) $F = AB\bar{C} + \bar{A}BC + A\bar{B}C\bar{D} + \bar{A}\bar{B}D$
c) $F = \bar{A}B\bar{C}D + A\bar{B}CD + \bar{A}\bar{B}C\bar{D}$

1.20 Map the expression $F = \Sigma(3, 8, 12, 14, 15)$. Obtain the minimal expressions for *a*) F, *b*) $\bar{F}$.

1.21 Simplify $F = \overline{(\overline{\bar{A}C + D})(\bar{A} + \bar{C})\bar{B}D}$ and $F = \overline{(AD + \overline{CD})(\overline{AC + B})}$

2 Logic Families

Introduction

The various kinds of gates described in Chapter 1 can be constructed in a number of different ways, some of which use discrete components while others use monolithic integrated circuit techniques. Discrete component logic is nowadays rarely, if ever, employed and it will not be considered in this chapter. The most popular logic families are *transistor–transistor logic* (ttl), of which several versions exist, and *complementary metal oxide semiconductor* (cmos), although *emitter-coupled logic* (ecl) is available for very fast applications. These families provide both *small-scale integrated* (ssi) devices, *medium-scale integrated* (msi) devices, *large-scale integrated* (lsi) devices, and *very-large-scale integrated* (vlsi) devices. The terms ssi and msi refer to devices having fewer than ten gates and fewer than 100 gates (or equivalent circuits), respectively. Similarly, lsi devices contain more than 100 gates and vlsi devices may contain several thousand gates.

Integrated circuits containing 1 to 4 gates are often referred to as *random logic* and have been the mainstay of most logic design in the past. In modern circuitry the use of msi and lsi devices such as multiplexers and microprocessors have led to fewer applications for random logic but it is expected to continue to be used for such purposes as interfacing to, from, and between the lsi devices. In addition, standard ssi devices can provide simple solutions to many digital requirements.

Some lsi devices employ other forms of logic that are not used for the simpler circuits; these are known as nmos, and integrated injection logic (I^2L). Lastly, some lsi circuits use a technique known as the charge-coupled device (ccd). All of these approaches to digital circuitry will be described in this chapter.

Electronic Switches

An **electronic switch** is a two-state device that has only two stable states, either the device is ON or it is OFF. The binary states 1 and 0 can be represented by an electronic switch; 1 can be represented by the OFF condition and 0 by the ON condition, or vice versa. Although a number of different devices are capable of operating as two-state circuits, usually either a semiconductor diode or a transistor (bipolar or field effect) is used.

The Semiconductor Diode

A **semiconductor diode** is able to operate as an electronic switch because it offers a low resistance to the flow of electric current in one direction and a high resistance in the other. When the diode is forward biased and conducts a relatively large current, it is said to be ON. When it is reverse biased and conducts only a very small current, it is said to be OFF. A generalized diode characteristic is given in Fig. 2.1(*a*). The diode conducts only when the forward bias voltage is greater than a threshold value labelled as V_f.

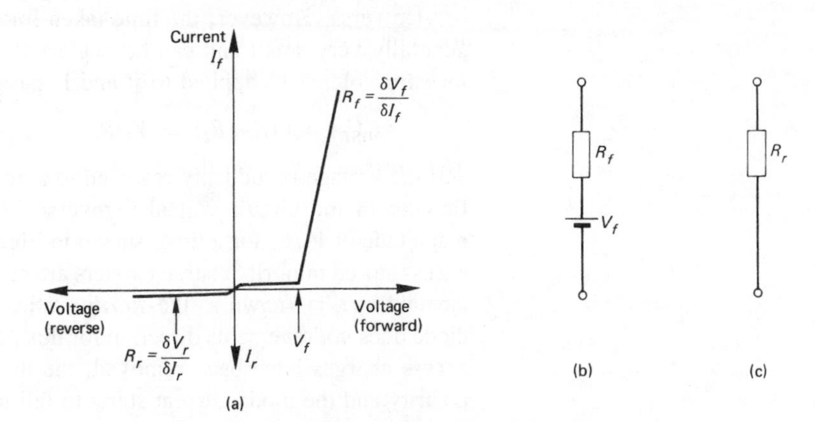

Fig. 2.1 (*a*) Generalized diode current/voltage characteristic (*b*) Equivalent circuit of a diode when conducting (*c*) Equivalent circuit of a diode when non-conducting

When a diode is ON it can be replaced, on paper, by an equivalent circuit consisting of a battery of e.m.f. V_f connected in series with a resistance R_f (see Fig. 2.1(*b*)).

V_f is the voltage dropped across the diode when it is conducting and it has a value that varies between about 0.4 V to 0.75 V depending upon the type of diode, the diode current, and the junction temperature. (A typical temperature coefficient is −2 mV/°C.) For silicon diodes V_f is very often assumed to be 0.6 V.

R_f is the a.c. resistance of the diode when it is forward biased. When a diode is OFF it can be represented by a resistance R_r (Fig. 2.1(*c*)), which is the reverse resistance of the diode.

The equivalent circuits of a diode are linear and they assist in the determination of the currents and voltages in a diode circuit.

Example 2.1

Calculate the output voltage of the circuit given in Fig. 2.2 when *a)* $V_1 = +5$ V, $V_2 = 0$ V, *b)* $V_1 = +5$ V, $V_2 = +5$ V, *c)* $V_1 = 0$ V, $V_2 = +5$ V, and *d)* $V_1 = 0$ V, $V_2 = 0$ V, if $V_f = 0.6$ V and $R_f = 10\ \Omega$.

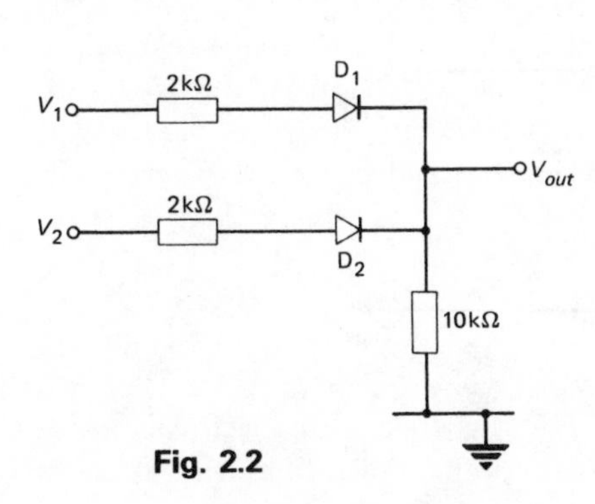

Fig. 2.2

Solution

a) Diode D_1 will be forward biased and ON while diode D_2 is OFF. Therefore

$$V_{out} = \frac{(5-0.6) \times 10\,000}{12\,010} = 3.66 \text{ V} \quad (Ans)$$

b) Both diodes are now ON and therefore

$$V_{out} = \frac{(5-0.6) \times 10\,000}{(\tfrac{1}{2} \times 2010) + 10\,000} = 4.0 \text{ V} \quad (Ans)$$

c) This is really the same situation as in *a*). Hence,

$$V_{out} = 3.66 \text{ V} \quad (Ans)$$

d) Neither diode conducts and so $V_{out} = 0$ V (*Ans*)

When the polarity of the voltage applied across a diode is reversed, the diode will not instantaneously change its state from ON to OFF or vice versa. However, the time taken for a diode to switch on is generally very small and can be neglected. Suppose a diode has a forward voltage V_F applied to it and is passing a current

$$I_F = (V_F - V_f)/(R_f + R_L) \simeq V_F/R_L$$

If the voltage is suddenly reversed to a new value V_R, the current flowing in the circuit will also reverse its direction and have a magnitude of V_R/R_L for a time, shown in Fig. 2.3 as t_s, during which excess stored minority charge carriers are removed. The time period labelled as t_s is known as the *storage time*. The voltage across the diode does *not* reverse its direction for this period of time. When the excess charges have been removed, the diode voltage reverses its polarity and the diode current starts to fall taking a time t_f to reach zero (Fig. 2.3).

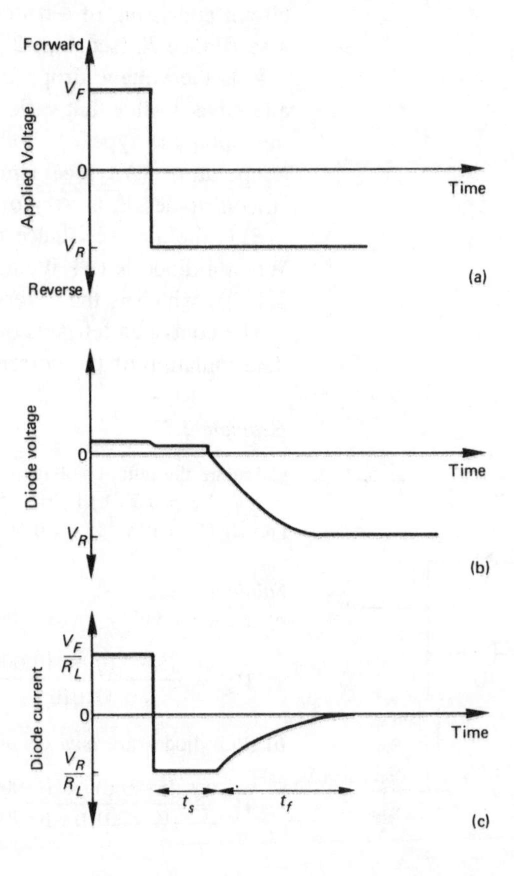

Fig. 2.3 Switching a diode: (*a*) applied voltage, (*b*) diode voltage, (*c*) diode current

Manfacturers' data generally quote the *reverse recovery time* of a diode. This is the time that elapses from the moment the diode current first reverses its direction of flow to the time the diode current reaches a defined value. The reverse recovery time may vary from a few nanoseconds to a few microseconds.

The Bipolar Transistor

Figure 2.4 shows a typical set of output characteristics for a **bipolar transistor** with a d.c. load line drawn between the points

$$V_{CE} = V_{CC} = 12 \text{ V}, I_C = 0 \text{ and } V_{CE} = 0 \text{ V}, I_C = V_{CC}/R_L = 6 \text{ mA}$$

where R_L is the total load on the transistor. When a transistor is used as switch it is rapidly switched between two stable states OFF and ON.

Fig. 2.4 The transistor as a switch

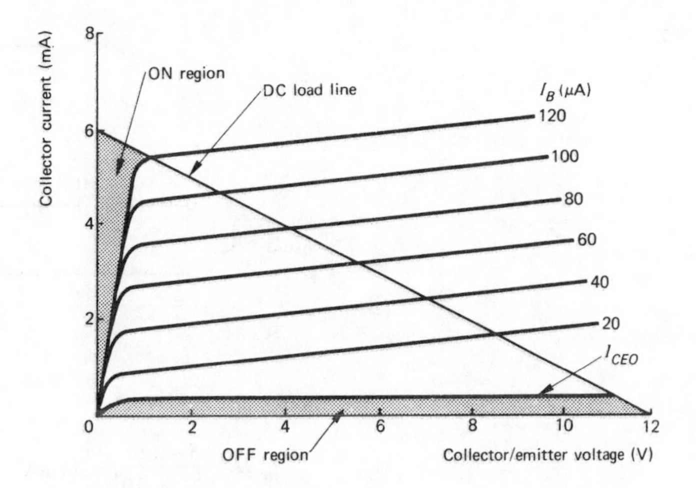

When the transistor is OFF, both its collector/base and emitter/base junctions are reverse biased and the collector current is only the small collector leakage current I_{CEO}. The collector/emitter voltage V_{CE} of the transistor is then equal to the supply voltage V_{CC}.

When the transistor is conducting current in its active region, the collector/base junction is reverse biased but the emitter/base junction is forward biased. As the base/emitter voltage V_{BE} is increased, the base current increases also and this produces an increase in the collector current since

$$I_C = h_{FE}I_B + I_{CEO} \simeq h_{FE}I_B$$

Eventually the point is reached at which the collector/base junction becomes forward biased and the transistor is said to be *saturated* or *bottomed*. The collector current now has its maximum, or ON, value $I_{C(sat)}$ and the base current is $I_B = I_{C(sat)}/h_{FE}$. The base/emitter voltage $V_{BE(sat)}$ producing this base current usually has a value of about

0.75 V. Any further increase in the base current will *not* produce a corresponding increase in the collector current. The collector/emitter voltage $V_{CE(sat)}$ of the transistor is then very low, being typically in the region of 0.1–0.2 V, because most of the supply voltage is dropped across the collector resistor R_L. Thus, the saturated collector current is equal to

$$I_{C(sat)} = \frac{V_{CC} - V_{CE(sat)}}{R_L} \simeq \frac{V_{CC}}{R_L} \quad (2.1)$$

Note that $V_{CC}/R_L > h_{FE}I_B$.

Example 2.2

The transistor shown in Fig. 2.5 has $h_{FE} = 50$. Determine whether or not the transistor saturates. If it does not find the maximum base resistance for saturation to occur. $V_{BE} = 0.6$ V and $V_{CE(sat)} = 0.2$ V.

Fig. 2.5

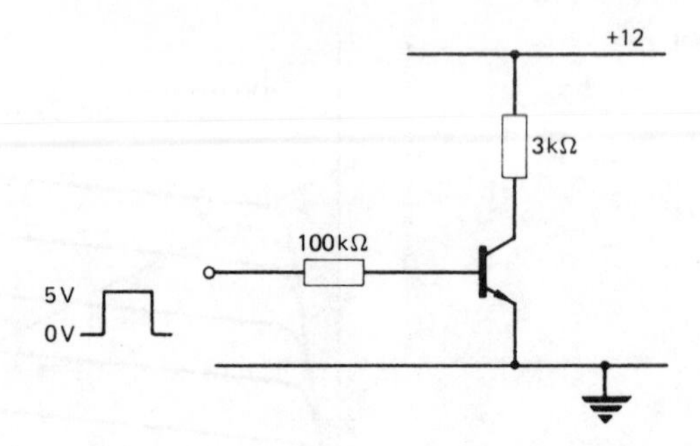

Solution

$$I_B = \frac{5-0.6}{100 \times 10^3} = 44\ \mu A$$

and so $I_C = 50 \times 44 \times 10^{-6} = 0.22$ mA.

When the transistor is saturated

$$I_{C(sat)} = \frac{12-0.2}{3 \times 10^3} = 3.93\ \text{mA}$$

For saturation to occur

$$I_B \geq \frac{3.93 \times 10^{-3}}{50} \geq 78.7\ \mu A$$

and so the transistor does *not* saturate.

For saturation to take place the maximum base resistance is equal to

$$(5-0.6)/78.7 \times 10^{-6} \simeq 56\ k\Omega \quad (Ans)$$

A transistor can be rapidly switched ON and OFF by the application of a rectangular waveform of sufficient amplitude to its base. In either of the two stable states the power dissipated within the transistor is small because *either* V_{CE} *or* I_C is approximately equal to zero. The active (or amplifying) region of the transistor is rapidly passed through as the transistor switches from one state to the other, and so little power is dissipated.

A transistor is unable to change state instantaneously when the voltage applied to its base is changed, because of charges stored in *a*) the base region, *b*) the collector/base depletion layer, and *c*) the base/emitter depletion layer. When the transistor is OFF both its base/emitter and its collector/base junctions are reverse biased and the two depletion layers are wide. When a voltage is applied to the base to turn the transistor ON, the base current supplies charge to both the p-n junctions and this reduces the widths of the two depletion layers. At some point the base/emitter depletion layer will be sufficiently narrow to allow charge carriers to move from the emitter into the base, and when these reach the collector a collector current starts to flow. As the collector current increases, the collector/emitter voltage falls and the width of the collector/base depletion layer decreases. If the base current is sufficiently large the collector current continues to increase until the transistor is saturated; the collector/base junction is then forward biased and so charge carriers are no longer swept into the collector region. This means that an *excess charge* is stored in the base region.

When a voltage is applied to the base to turn the transistor OFF, the collector current will not start to fall until all the excess base charge has been removed. This time delay is known as the *storage delay*.

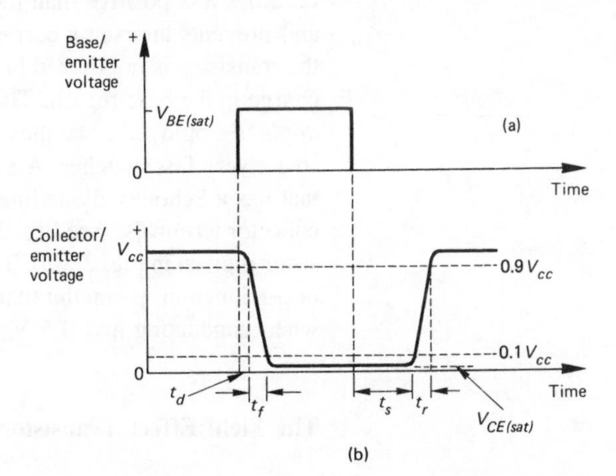

Fig. 2.6 Switching a bipolar transistor circuit: (*a*) input voltage, (*b*) output voltage

Figure 2.6 shows how the collector/emitter voltage of an initially OFF bipolar transistor varies when a voltage pulse is applied to its base. The terms shown in the figure are defined thus:

t_d is the time that elapses between the application of the base signal and the collector/emitter voltage falling to 90% of its original value of V_{CC} volts.

t_f is the time taken for the collector/emitter voltage to fall from 90% to 10% of V_{CC} volts.

t_s is the time delay that occurs between the removal of the base voltage and the collector/emitter voltage rising to 10% of its final value of V_{CC} volts.

t_r is the time taken for the collector/emitter voltage to rise from 0.1 V_{CC} to 0.9 V_{CC} volts.

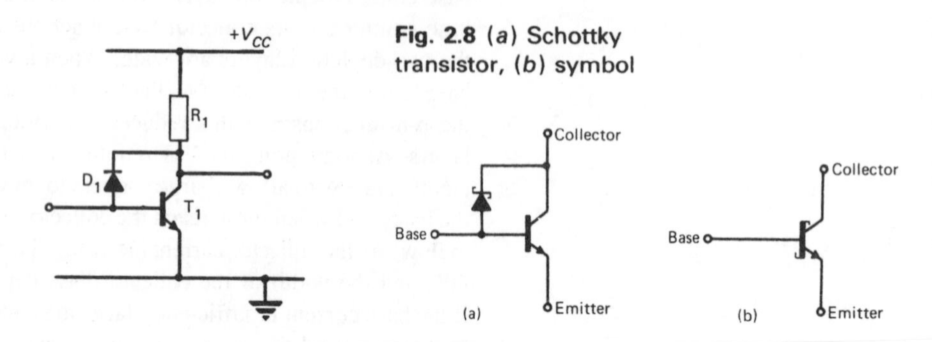

Fig. 2.7 Use of a diode to increase switching speed

Fig. 2.8 (*a*) Schottky transistor, (*b*) symbol

Typically, the ON and OFF times are about 6 ns and 10 ns respectively, and to increase the switching speed the transistor must be prevented from saturating. This can be achieved by the connection of a diode between the base and the collector terminals of the transistor as shown in Fig. 2.7. When the transistor is turned ON, its collector/emitter voltage falls, and when the collector potential becomes less positive than the base potential the diode D_1 conducts and prevents an excess current from entering the base. As a result the transistor is not driven into saturation and there is no storage of charge in the base region. The best results are obtained if a *Schottky diode* is employed since these devices have zero charge storage and so are very fast switches. A Schottky transistor is a bipolar transistor that has a Schottky diode internally connected between its base and collector terminals, (see Fig. 2.8(*a*)). The symbol for a Schottky transistor is given in Fig. 2.8(*b*). The voltage drop across a Schottky diode, or p-n junction, is smaller than for an ordinary diode, typically 0.3 V when conducting and 0.5 V when full ON.

The Field Effect Transistor

A **field effect transistor** can also be employed as an electronic switch since its drain current can be turned ON and OFF by the application of a suitable gate-source voltage. When the fet is ON, the gate-source voltage will have moved the operating point to the top of the load line (similar to the bipolar transistor switch), and the voltage across the fet, known as the saturation voltage $V_{DS(SAT)}$, is small, typically 0.2 V–1.0 V.

When OFF the fet passes a very small current, typically 1 nA for

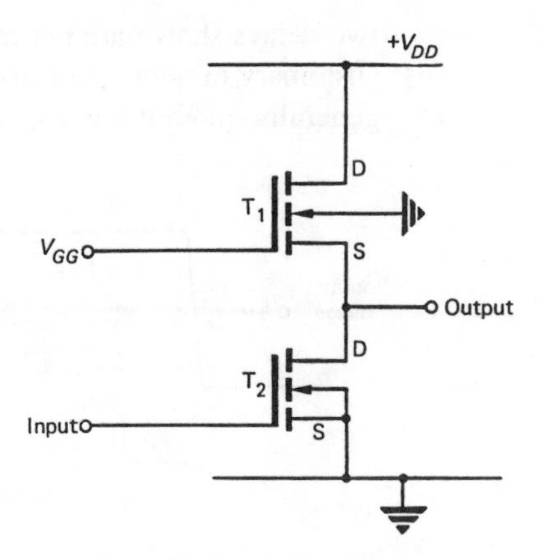

Fig. 2.9 Enhancement mosfet switch; T_1 acts as an active load for T_2

a junction fet and 50 pA for a mosfet. The drain load resistance needed for a mosfet switch is often of the order of some tens of kilohms and such high values are not conveniently fabricated within a monolithic integrated circuit. For this reason the drain load is often provided by another mosfet as shown in Fig. 2.9. The bottom transistor T_2 is operated as the switch while the upper transistor T_1 is biased to act as an *active load* by the voltage V_{GG}, where $V_{GG} > V_{DD}$. Very often $V_{GG} = V_{DD}$ and a separate bias voltage supply is then not necessary. The active load transistor T_1 is always conducting current and often has its substrate connected to its source instead of to earth.

The switching speed of a fet is determined by the stray and transistor capacitances which are unavoidably present in the circuit. Because of the very high input impedance of a mosfet, the time constant of its input circuit is a speed-limiting factor. The charge storage effects encountered with the bipolar transistor circuit are now insignificant because in a fet current is carried by the majority charge carriers.

Parameters of the Logic Families

The various logic families that will be discussed in this chapter possess different characteristics, which means that any one of them may be the best suited for a particular application. For example, for one application the most important consideration might be the highest possible speed of operation, whilst for another application it might be the minimum possible power dissipation. The characteristics of the various logic families can be classified under the following headings: speed of operation, fan-in and fan-out, noise margin or noise immunity, and power dissipation.

1 Speed of Operation The speed of operation, or the propagation delay, of a logic circuit is the time that elapses between the application of an input signal and the resulting change in the logic state at the output (see Fig. 2.10). Generally, the propagation delay is measured at the 50% points on the input and output waveforms. The two delays shown are not necessarily equal to one another and it is customary to quote their average value. Manufacturers' data sheets generally quote the worst-case propagation delay.

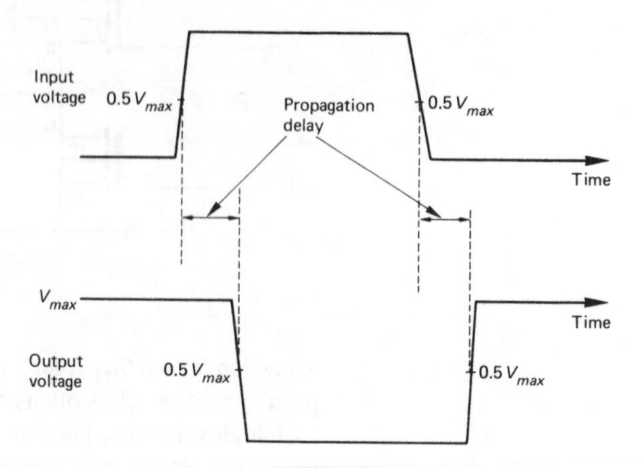

Fig. 2.10 Speed of operation of a logic device

2 Noise Margin or Noise Immunity The false operation of a logic circuit can be caused by transient voltages produced by switching, and by noise voltages induced from other logic lines and/or generated within the power supplies. If the noise voltage is of sufficiently high value, it may cause a gate to change its output state even though the input signal voltage has remained constant. The noise margin or noise immunity of a gate is the maximum noise voltage that can appear at its input terminals without causing a change in the output state.

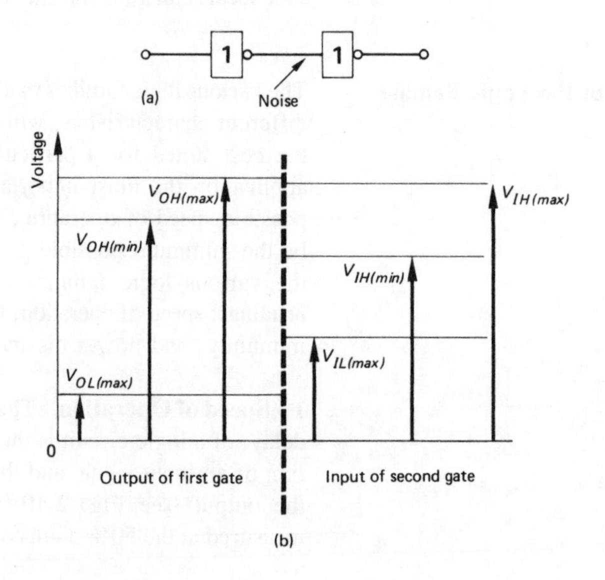

Fig. 2.11 Noise margin

Consider as an example the case of two cascaded NAND or NOR gates shown in Fig. 2.11(*a*). Suppose that the gates have a range of possible high-level and low-level input voltages as shown in Fig. 2.11(*b*). The output voltage of a gate may vary between limits $V_{OH(max)}$ and $V_{OH(min)}$ for the logical 1 output and between the limits $V_{OL(max)}$ and $V_{OL(min)}$ for the output to be at logical 0 (assuming positive logic). Similarly, the input voltage of a gate may vary between limits $V_{IH(max)}$ and $V_{IH(min)}$ for the input state to be at logical 1. For the input state to be at logical 0 the input voltage must be between the limits $V_{IL(max)}$ and $V_{IL(min)}$. The noise margin of a gate will then be

$$V_{OH(min)} - V_{IH(min)} \quad \text{and} \quad V_{IL(max)} - V_{OL(max)}$$

When noise margins are quoted in manufacturers' data sheets it is usual to give the worst-case values; for example the worst-case values for the logic voltages of a ttl NAND gate are

$$V_{OH(min)} = 2.4 \text{ V} \qquad V_{OL(max)} = 0.4 \text{ V}$$
$$V_{IH(min)} = 2.0 \text{ V} \qquad V_{IL(max)} = 0.8 \text{ V}$$

so that the worst-case noise margins are

$$2.4 - 2.0 = 0.4 \text{ V} \quad \text{and} \quad 0.8 - 0.4 = 0.4 \text{ V}$$

This is the guaranteed noise margin for ttl devices but it is typically at least 1 V.

3 Power Dissipation Power is dissipated within a transistor as it switches from one state to another and also within all current-carrying resistors. The d.c. power dissipation of a gate is the product of the d.c. supply voltage and the mean current taken from that supply. Typical figures are quoted in data sheets.

4 Fan-in and Fan-out The **fan-in** of a gate is the number of inputs connected to the gate. The **fan-out** of a gate is the maximum number of standard loads that can be connected to its output terminals without the output voltage falling outside the limits at which the logic levels 0 and 1 are specified.

A *standard load* is the load provided by a single simple input stage; for ttl circuits this is 1.6 mA. Some of the more complex circuits may be equivalent to two, or more, standard loads.

Current Sinking and Current Sourcing

Current sinking and **current sourcing** are two terms commonly employed in digital work to describe the basic operation of a logic circuit.

A **current sink** is a circuit that is supplied with a current by another circuit, while a **current source** supplies a current to a sink load. The difference in meaning between the two terms is illustrated by Fig.

2.12. In Fig. 2.12(*a*) current will flow *into* the sink circuit when the switch is closed, while in Fig. 2.12(*b*) closure of the switch makes a current flow *out* of the source circuit.

For any logic circuit a sink load is one that goes to earth via a forward-biased p-n junction (diode or the base/emitter junction of a transistor) and tends to reduce the logical 1 level of the circuit driving it (see Fig. 2.12(*c*)). Conversely, a source load is one that goes to V_{CC} volts via a forward-biased junction (Fig. 2.12(*d*)) and tends to increase the logical 0 output voltage level of the previous circuit.

Sometimes a circuit can be *either* a sink *or* a source depending upon its operating conditions. Consider Fig. 2.13 (which will later be recognized as a totem-pole output stage for a ttl gate). When T_1 is OFF and T_2 is ON, the output terminal is connected to earth via T_2. This allows a positive supply voltage connected via a forward-biased p-n junction to the output terminal to pass a current I_{OL} via T_2 to earth. Hence T_2 acts as a current sink. Conversely, if T_1 is ON and T_2 is OFF, the output terminal is connected to the positive supply line $+V_{CC}$ via T_1. If, therefore, the output terminal is connected, via a forward-biased p-n junction, to earth, a current will flow from V_{CC} via T_1 to earth so that T_1 acts as a current source.

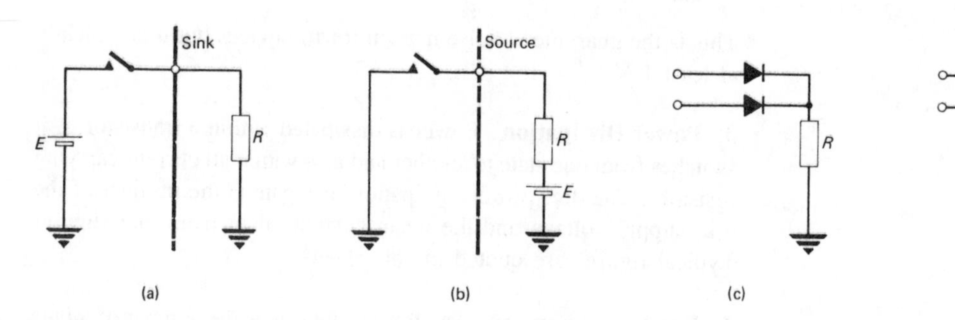

Fig. 2.12 Current sinking and sourcing, (*a*) and (*c*) are current sinks, (*b*) and (*d*) are current sources.

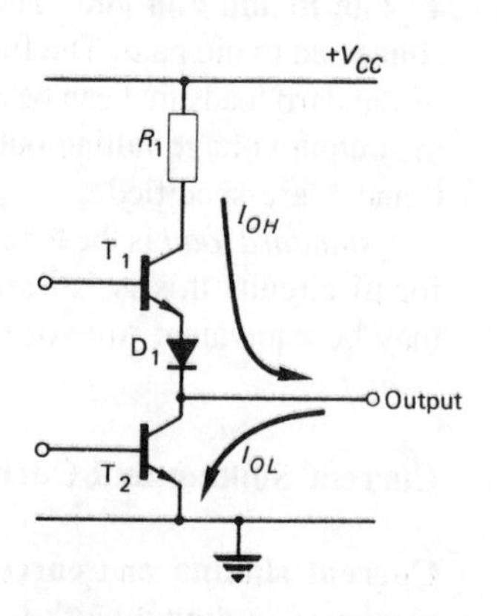

Fig. 2.13 A circuit that can be either a current sink or a current source

Logic Levels

Ideally, the logical 0 and 1 voltage levels are 0 V and V_{CC} volts respectively but, in practice, both these levels will differ from the ideal. Consider Fig. 2.14 which shows a transistor that represents the output stage of a gate. When the transistor is ON the voltage across it—the logic 0 voltage level—is $V_{CE(SAT)}$ or approximately 0.2 V. The logical 1 voltage level will depend upon the number N of similar gates (the fan-out) that are connected in parallel across the output terminals. The total resistance across the output terminals is R_{IN}/N where R_{IN} is the input resistance of each of the N identical gates.

Fig. 2.14 Logic levels

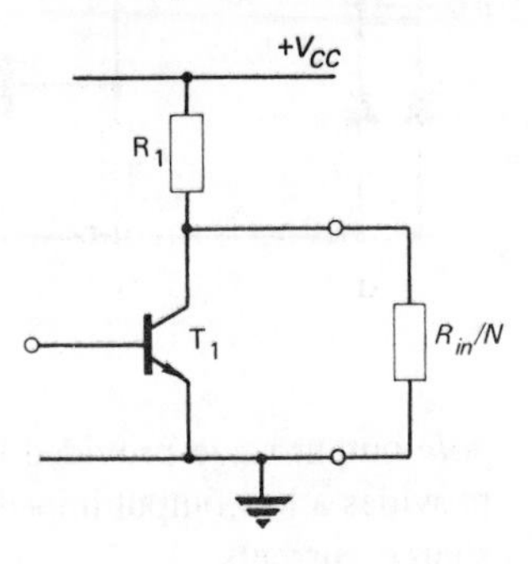

The output 1 voltage level is

$$\frac{V_{CC}\, R_{IN}/N}{R_1 + R_{IN}/N}$$

and it is clear that the greater the fan-out, the lower the logical 1 voltage level. The maximum fan-out is limited by the allowable minimum voltage specified for the logical 1 state; for ttl, for example, this is 2.2 V.

Transistor–Transistor Logic

The most popular and widely used logic family is the **transistor–transistor logic** or **ttl** family. TTL is manufactured in integrated circuit form by several manufacturers. The great popularity of this logic family arises because it can offer high speed, particularly the advanced versions, good fan-in and fan-out, and it is easily interfaced with other digital circuitry. In addition, it is cheap and readily available from many sources and several versions exist.

Standard ttl

The standard ttl logic, known as the 54/74 series, has a poor noise immunity and a rather high power consumption. The 74 series is designed for commercial applications and can operate at ambient temperatures of up to 70 °C. The 54 series is primarily intended for military applications and has a maximum ambient temperature figure of 125 °C.

The circuit of a standard **ttl NAND gate** is shown in Fig. 2.15. Only two inputs are shown but the fan-in may be up to 8. The *totem*

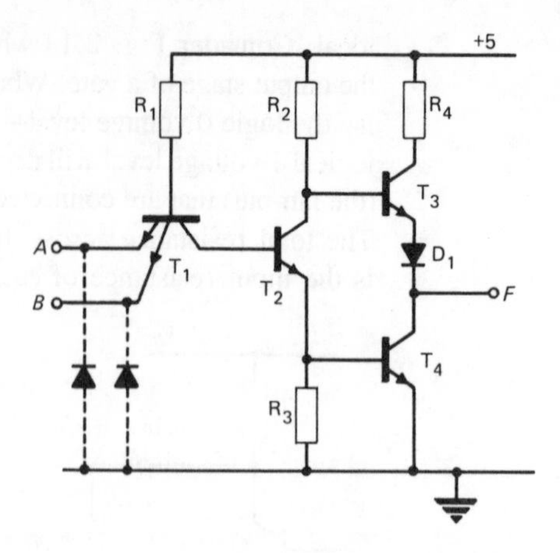

Fig. 2.15 Standard ttl NAND gate

pole output stage provided by transistors T_3 and T_4 is used since it provides a low output impedance and the capability to both sink and source currents.

When both input terminals are at logical 1 ($\simeq +5$ V) the emitter/base junctions of the multiple-emitter transistor T_1 are reverse biased *but* its collector/base junction is forward biased. This means that the operation of T_1 is *inverted* and its current gain h_{FE1} is less than unity. Current flows from the collector power supply through R_1 and T_1 into the base of transistor T_2. The base current of T_2 is $-(1 + h_{FE1})I_{B1}$ which is large enough to drive T_2 into saturation and so the collector/emitter voltage of T_2 is approximately 0.2 V.

The base/emitter voltage V_{BE4} of T_4 is the voltage developed across R_3 and this turns T_4 ON so that the output voltage falls to $V_{CE4(SAT)}$ or about 0.2 V. The base/emitter voltage of T_3 is now equal to

$$V_{CE2(SAT)} + V_{BE4(SAT)} - (V_{CE4(SAT)} + V_{D1})$$

$$= 0.2 + 0.75 - (0.2 + 0.6) = 0.95 \text{ V} - 0.8 \text{ V} = 0.15 \text{ V}$$

and so this transistor turns OFF. Thus, the logical state of the output is 0 when both the inputs are at logical 1 and so the NAND function is performed. If the diode D_1 were not present the base/emitter voltage V_{BE3} of T_3 would be equal to $0.95 - 0.2 = 0.75$ V and this would be large enough for T_3 to conduct. With D_1 in circuit

$$V_{BE3} = 0.95 - 0.6 - 0.2 = 0.15 \text{ V}$$

When one or more of the inputs is at logical 0 (approximately 0.2 V), the associated emitter/base junction(s) of T_1 is forward biased and T_1 is turned fully ON. The collector potential of T_1 is then

$$V_{CE1(SAT)} + 0.2 = 0.2 + 0.2 = 0.4 \text{ V}$$

This potential is not large enough to keep T_2 conducting and so T_2 turns OFF. The collector potential of T_2 is now $+5$ V and its emitter potential is 0 V and T_3 turns ON and T_4 turns OFF. The output voltage does not immediately change its value because of the stray capacitance across the output terminals. T_3 sources current to this capacitance and the output voltage rises exponentially towards + 5 V. As the charging current falls, T_4 comes out of saturation and V_{out} attains a steady value equal to

$$V_{CC} - V_{BE3} - V_{D1}$$

Example 2.3

In the circuit of Fig. 2.15, $R_1 = 4\ k\Omega$, $R_2 = 1.4\ k\Omega$, $R_3 = 1\ k\Omega$, $R_4 = 100\ \Omega$. If, for each transistor $V_{CE(SAT)} = 0.2$ V and $V_{BE(SAT)} = 0.6$ V, calculate *a*) the minimum value of h_{FE} for T_3 to change state, *b*) the output voltage when one or more of the inputs is at the logical 0 level, and *c*) the output voltage when all the inputs are high.

Solution

a) $V_{B3} = V_{D1} + V_{out} + V_{BE3} = 1.4$ V

Therefore $I_{B3} = (V_{CC} - V_{BE3})/1.4 = 2.57$ mA

$$I_{C3} = (V_{CC} - V_{CE(SAT)} - V_{D1} - V_{out})/R_4$$
$$= (5 - 0.2 - 0.6 - 0.2)/100 = 40 \text{ mA}$$

Hence $h_{FE(min)} = 40/2.57 = 15.6$ (*Ans*)

b) Steady-state output voltage is

$$V_{CC} - V_{BE3} - V_{D1} = 5 - 0.6 - 0.6 = 3.8 \text{ V} \quad (Ans)$$

c) When T_1 is OFF, T_4 is ON. Therefore

Output voltage $V_{CE(SAT)} = 0.2$ V (*Ans*)

Example 2.4

The circuit of Fig. 2.15 has $R_1 = 4\ k\Omega$, $R_2 = 1.6\ k\Omega$, $R_3 = 1\ k\Omega$, $R_4 = 130\ \Omega$, $h_{FE1} = 1$ and $h_{FE2} = h_{FE3} = h_{FE4} = 30$. Calculate the currents and voltages in the circuit.

Solution

When all the inputs are at logical 1,

$$I_{C1} = (1 + h_{FE1})I_{B1} = 2I_{B1} = I_{B2}$$

T_2 and T_4 are saturated and T_3 is OFF because its base voltage is 0.2 + 0.75 = 0.95 V, while its emitter voltage is 0.2 + 0.6 = 0.8 V. The base potential of T_1 is

$$V_{BE4(SAT)} + V_{BE2(SAT)} + V_{CB1} = 0.7 + 0.7 + 0.7 = 2.1 \text{ V}$$

and hence

$$I_{B1} = (5 - 2.1)/4 \times 10^3 = 0.725 \text{ mA}$$

Also $I_{B2} = 2I_{B1} = 1.45$ mA

The collector current of T_2 is

$$(V_{CC} - V_{BE3})/R_2 = (5 - 0.95)/(1.6 \times 10^3) = 2.53 \text{ mA}$$

The current flowing in R_3 is $0.7/10^3 = 0.7$ mA.
The emitter current of T_2 is

$$I_{E2} = I_{B2} + I_{C2} = 1.45 + 2.53 = 3.98 \text{ mA}$$

Hence $I_{B4} = 3.98 - 0.7 = 3.28$ mA

T_3 is OFF, hence $I_{C3} = 0$. I_{C4} depends upon the load connected across the output terminals, i.e. upon the fan-out.

If any one or more of the inputs is at logical 0, or approximately 0.2 V, then T_2 and T_4 turn OFF and T_3 saturates. The base potential of T_3 is initially at

$$V_{BE3(SAT)} + V_{D1} + V_{out} = 0.75 + 0.7 + 0.2 = 1.65 \text{ V}$$

and so

$$I_{B3} = (5 - 1.65)/1.6 \times 10^3 \simeq 2.1 \text{ mA}$$

Also

$$I_{C3} = (V_{CC} - V_{CE3(SAT)} - V_{D1} - V_{out})/R_4$$
$$= (5 - 0.2 - 0.7 - 0.2)/130 = 30 \text{ mA}$$

Note that this means that h_{FE3} must be at least equal to $30/2.1 = 14.3$.

The output voltage rises exponentially towards 5 V and settles down at a steady value of

$$5 - V_{BE3} - V_{D1} = 5 - 0.6 - 0.6 = 3.8 \text{ V}$$

Fan-out

The **fan-out** is limited by the current that T_4 can sink when it is saturated. Referring to Fig. 2.16 with one gate only connected to the output terminals,

$$I_{out} = (V_{CC} - V_{BE1} - V_{out})/R_1$$
$$= (5 - 0.75 - 0.2)/4000 = 1.01 \text{ mA}$$

For a fan-out of N the output current will be 1.01 N mA. For T_4 to

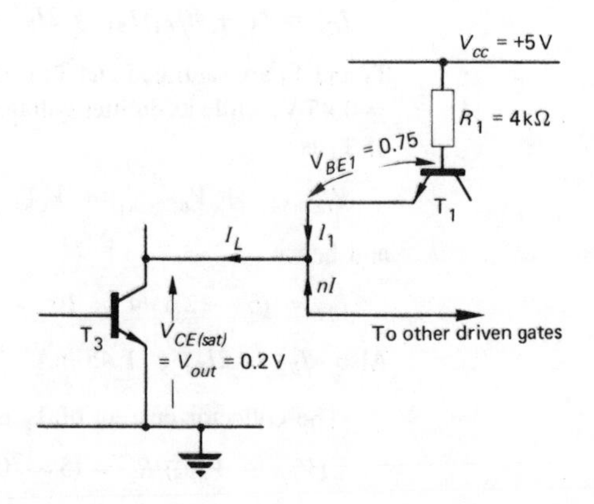

Fig. 2.16 Calculation of fan-out

be in saturation, $I_{B4}h_{FE4} > 1.01\ N$ and so the maximum fan-out is given by

$$I_{B4}h_{FE4}/1.01$$

There is, however, another factor to be considered and that is the maximum output voltage that can represent logic 0. The standard value for ttl is 0.4 V and, to keep V_{out} below this, the fan-out is normally restricted to ten.

Noise Margins

When all inputs are at logical 1 or +3.8 V, the base potential of T_1 is 0.7 + 0.7 + 0.7 = 2.1 V. For a diode to remain in its reverse-biased state when a noise voltage V_n is present, then

$$2.1 - 3.8 - V_n \leq 0.7 \quad \text{or} \quad V_n \geq 2.1 - 3.8 - 0.7 \geq -2.4 \text{ V}$$

Similarly, when any one of the inputs is at logical 0, the base potential of T_1 is 0.2 + 0.2 = 0.4 V. A noise voltage V_n will cause false operation of the circuit if

$$0.4 + V_n \geq V_{BE2} + V_{BE4} \geq 0.7 + 0.7 \geq 1.4 \text{ V}$$

Hence the noise margin is 1.4 − 0.4 = 1.0 V.

The *worst-case* noise margins are equal at 400 mV.

Voltage Levels

For a ttl NAND gate, the input and output voltages required to specify the 1 and 0 logical states are

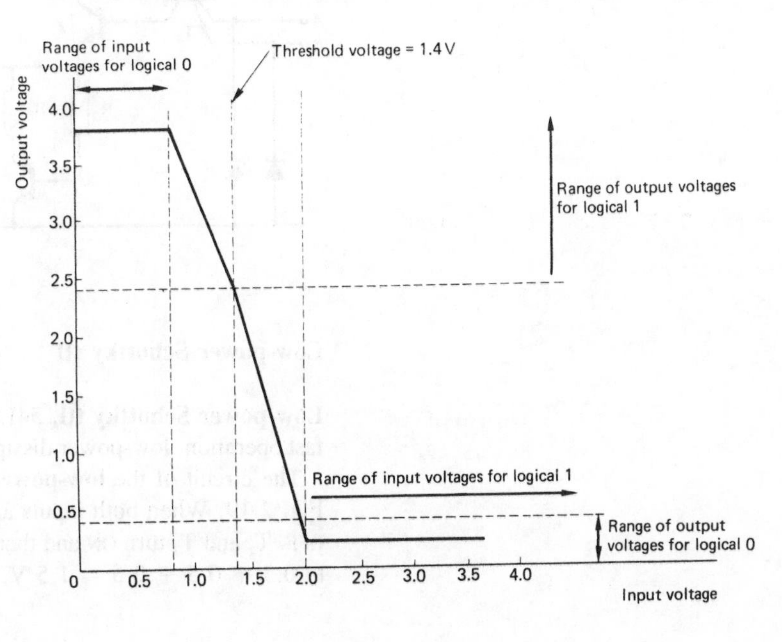

Fig. 2.17 TTL transfer characteristic

$$V_{01(min)} = 4.4 \text{ V} \qquad V_{00(max)} = 0.4 \text{ V}$$

$$V_{11(\min)} = 2.0 \text{ V} \qquad V_{10(max)} = 0.8 \text{ V}$$

The sink current $I_{OL} = 16$ mA and the source current $I_{OH} = 0.8$ mA. These figures can be expressed by the *transfer characteristic* of the gate shown in Fig. 2.17.

The *threshold voltage* is the input voltage at which a change of the output state of the gate is just triggered. A reasonable approximation to this value is the voltage midway between the $V_{10(max)}$ and $V_{11(min)}$ values, i.e. 1.4 V.

Schottky ttl

A decrease in the propagation delay of a gate can be achieved if Schottky diodes and transistors are used and a Schottky ttl logic series, 54S/74S, is available from several sources. The circuit of the NAND gate is given in Fig. 2.18 and it can be seen to be similar to the standard ttl NAND gate except that *a*) resistor R_2 is replaced by transistor T_4 and R_3/R_5 and *b*) diode D_1 is replaced by T_3/T_5 to increase the speed of operation.

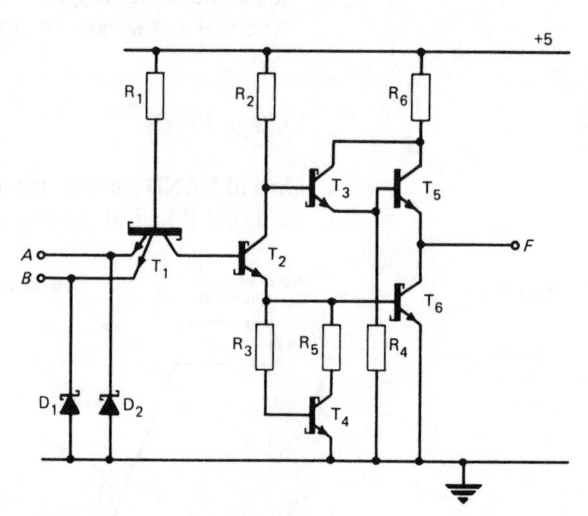

Figure 2.18 Schottky ttl NAND gate
$R_1 = 2.8$ k
$R_2 = 900\ \Omega$
$R_3 = 500\ \Omega$
$R_4 = 3.5$ k
$R_5 = 250\ \Omega$
$R_6 = 50\ \Omega$

Low-power Schottky ttl

Low-power Schottky ttl, 54LS/74LS, combines the advantages of fast operation, low-power dissipation, and high-frequency capability.

The circuit of the **low-power Schottky NAND gate** is shown in Fig. 2.19. When both inputs are at logical 1, diodes D_1 and D_2 are OFF. T_1 and T_3 turn ON and then the voltage at the base of T_1 is equal to $0.5 + 0.5 + 0.5 = 1.5$ V. The collector potential of T_3 is low

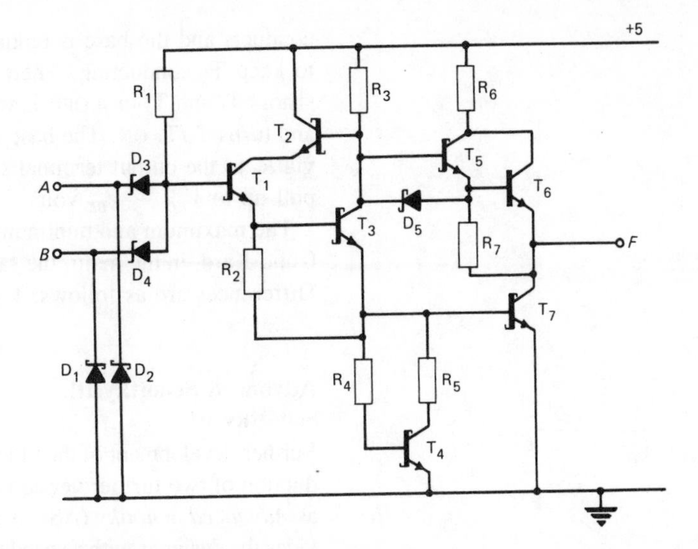

Fig. 2.19 Low-power Schottky ttl NAND gate

and so the Darlington pair T_5/T_6 [EIV] turns OFF. When the base/emitter voltage of T_3 is very nearly equal to $V_{BE(SAT)}$, then T_4 will be conducting and will take most of the emitter current of T_3. Transistor T_7 will not receive sufficient base current to turn it ON until the base potential of T_1 is high enough for T_1 to supply sufficient current. T_7 then rapidly turns ON. This has the effect of squaring the shape of the transfer characteristic and thereby increasing the noise immunity of the circuit (Fig. 2.20).

The speed of switching is increased by T_2 and D_5 whereby T_2 supplies a current surge during the turn-on of T_1 while D_5 provides a discharge path for T_6.

When one or more inputs are at logical 0, the associated input diode

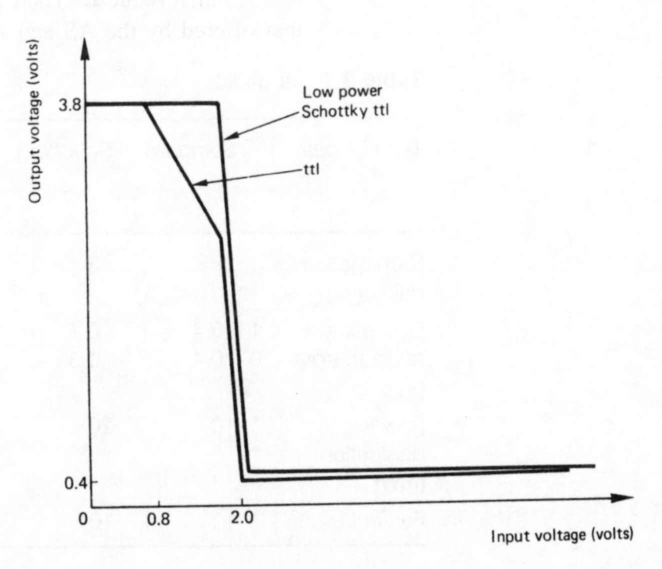

Fig. 2.20 Transfer characteristics of ttl and low-power Schottky ttl gates

conducts and the base potential of T_1 falls below the value needed to keep T_1 conducting. Therefore T_1 turns OFF. This makes transistors T_3 and T_7 turn OFF as well. The collector potential of T_3 rises and turns T_5/T_6 ON. The base of the output transistor T_6 is returned via R_7 to the output terminal since this allows the output voltage to pull up to $V_{CC} - V_{BE}$ volts.

The maximum and minimum input and output voltages for logical 0 and 1 are, in the main, the same as those for the standard ttl gates. Differences are as follows: $V_{OH(min)} = 2.7$ V and $V_{OL(max)} = 0.5$ V.

Advanced Schottky ttl

Further development of the ttl logic family has brought about the introduction of two further versions of Schottky ttl. One of these, known as *Advanced Schottky* (AS), is a development of Schottky ttl. It provides the designer with a speed performance which is comparable with that given by the ecl family of devices. The other version, known as *Advanced Low-power Schottky* (ALS), is an improved version of the low-power Schottky logic family giving both increased speed and a reduction in power dissipation. The speed of operation of ALS is, however, less than that of AS.

A large number of ssi and msi circuits are presently available in both versions of Advanced Schottky, and further circuits are continually being added to the catalogues. In general, the circuits offered correspond to ones which are readily available in the standard and LS branches of the ttl family.

An alternative development of the Schottky ttl family is known as FAST (Fairchild Advanced Schottky ttl). Circuits in this family offer the same speed of operation as Schottky ttl but their power dissipation is much reduced. Their performance is somewhere in between that offered by the AS and ALS families.

Table 2.1 ttl gates

Type of gate		*Standard*	*Schottky*	*Low-power Schottky*	*Advanced low-power Schottky*	*Advanced Schottky*
Propagation delay (ns)		9	3	7	4	1.5
Guaranteed noise margin (V)	1	0.4	0.7	0.7	0.7	
	0	0.4	0.3	0.3	0.4	
Power dissipation (mW)		10	20	2	1	2
Fan-out		10	10	10	10	10

All three versions of advanced Schottky ttl are now produced by several different manufacturers.

A comparison between the various types of ttl gates is given by Table 2.1 which shows typical values for the parameters quoted.

CMOS Logic

The **complementary metal-oxide semiconductor** or **cmos** logic family offers some significant advantages over bipolar logic such as very low power dissipation, good noise immunity, and the ability to operate from a wide range of power supply voltages (2–15 V). Its main disadvantages are its relatively long propagation delay and its low output current capability. The long propagation time arises from the input time constant of the enhancement-mode mosfets which are employed because they turn off when their gate-source voltage is at zero volts.

The circuit of a cmos inverter is given in Fig. 2.21. When the input A is high, the n-channel mosfet T_2 turns ON and the p-channel mosfet T_1 turns OFF. Hence the output of the circuit is low. Conversely, when the input A is low, T_1 turns ON and T_2 turns OFF and the output is then high.

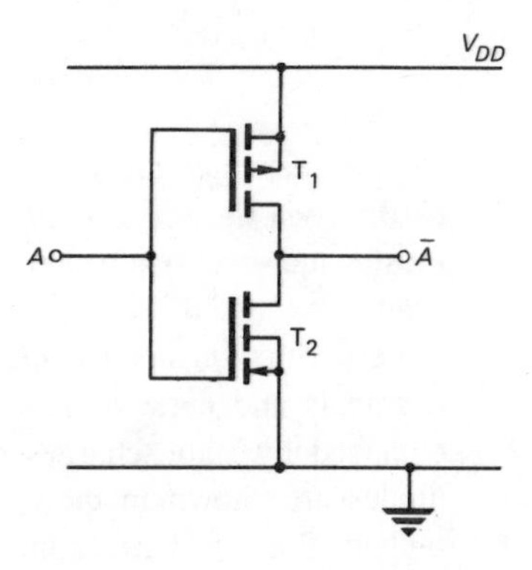

Fig. 2.21 CMOS inverter

The power dissipation of the circuit is very small because the only path between the $+V_{DD}$ supply and earth consists of an n-channel mosfet and a p-channel mosfet connected in series, and one or other of these is always turned OFF. Power is dissipated in the circuit only while the circuit is actually switching from one state to the other.

The circuit of a **cmos NAND gate** is given in Fig. 2.22. It can be seen that the n-channel mosfets are connected in series and the p-channel devices are in parallel. The operation of the circuit is as follows. If either, or both, of the inputs is at logical 0 ($\simeq$ 0 V), then the associated p-channel mosfets, T_1 and/or T_2, are turned ON, while the associated n-channel mosfets, T_3 and/or T_4, are turned OFF. The output terminal of the circuit is then at + 5 V minus the saturation voltage of an ON mosfet. Conversely, if both inputs are at logical 1 ($\simeq$ 5 V), T_1 and/or T_2 are turned OFF, and T_3 and/or T_4 are turned ON. The output of the circuit is then at approximately 0 V or logical 0.

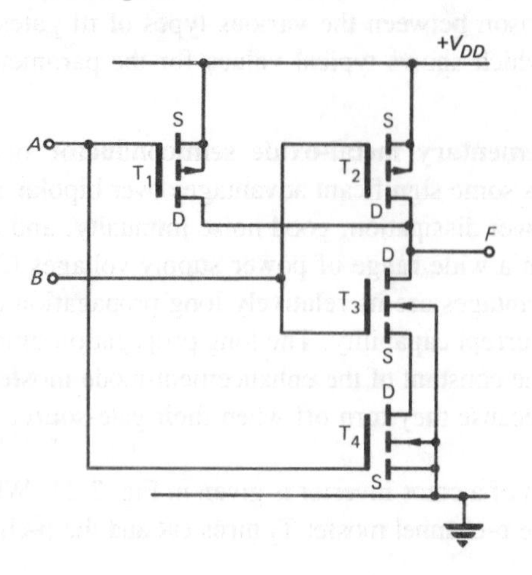

Fig. 2.22 CMOS NAND gate

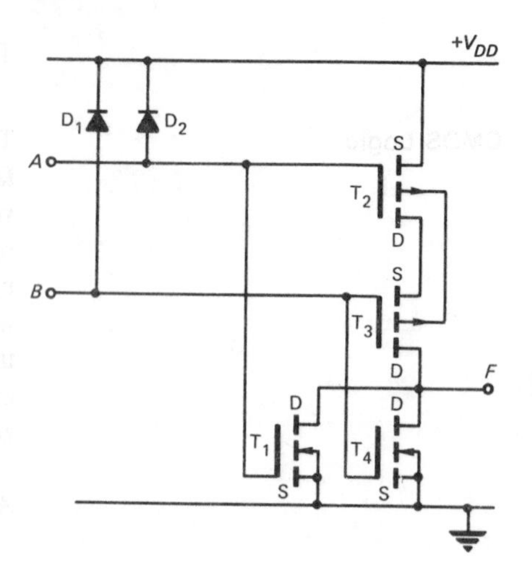

Fig. 2.23 CMOS NOR gate

Note that once again the only path between the $+V_{DD}$ line and earth is via the series connection of an n-channel and a p-channel mosfet and so zero power is dissipated when the gate is in either logic state.

Very often *protective diodes* are connected between the input terminals and earth to reduce the possibility of the device being damaged by static charges produced by handling. The protective diodes are shown in the circuit of Fig. 2.23. Note that now the p-channel mosfets are connected in series and the n-channel mosfets are connected in parallel. If either of the input terminals is at logical 1, the associated p-channel mosfet (T_2 and/or T_3) is turned OFF and the associated n-channel mosfets (T_1 and/or T_4) are turned ON. The output voltage of the circuit is then low so that the NOR function is performed. Only if both inputs are at logical 0 will both the n-channel devices turn OFF and the p-channel mosfets turn ON so that the output can reach its logical 1 state.

Another useful circuit available in the cmos family is known as the transmission gate and Fig. 2.24 shows the basic circuit arrangement. Assume that the clock C is at the logical 1 voltage level. Then, if $A = 1$, the gate-source voltage V_{GS1} of fet T_1 is zero and so T_1 will be OFF. At the same time, $\bar{C} = 0$ and so the gate-source voltage V_{GS2} of T_2 is negative and T_2 is ON. If $A = 0$, V_{GS1} is positive and T_1 is ON while V_{GS2} is zero and T_2 is OFF.

Conversely, when the clock C is at the 0 logic level (and $\bar{C}$ is 1), both the mosfets are turned OFF regardless of the logic level of the input A. Thus, the operation of the transmission gate is that, when the clock is 1, the output F of the circuit is equal to the input A but, when the clock is 0, there is no transmission through the circuit.

Fig. 2.24 CMOS transmission gate

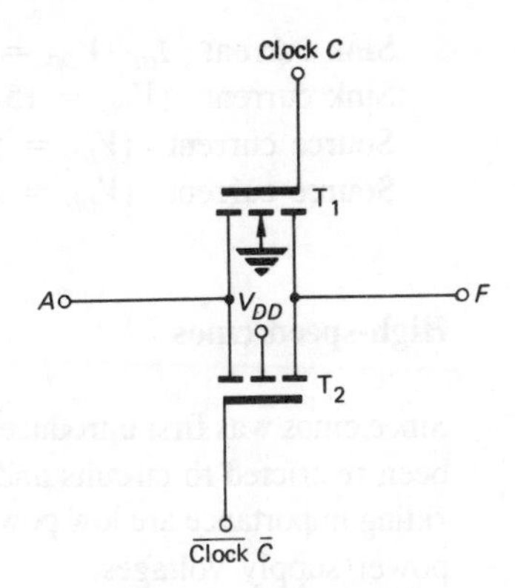

The necessary inversion of the clock signal is generally provided within the *transmission gate* circuit in the way shown by Fig. 2.25. The transistors T_1 and T_2 act as switches that can be turned ON (shut) or OFF (open) by a control signal applied to their gate terminals. One of the gate voltages should be the inverse of the other but this requirement is satisfied by means of the inverter provided by transistors T_3 and T_4.

Fig. 2.25 Transmission gate

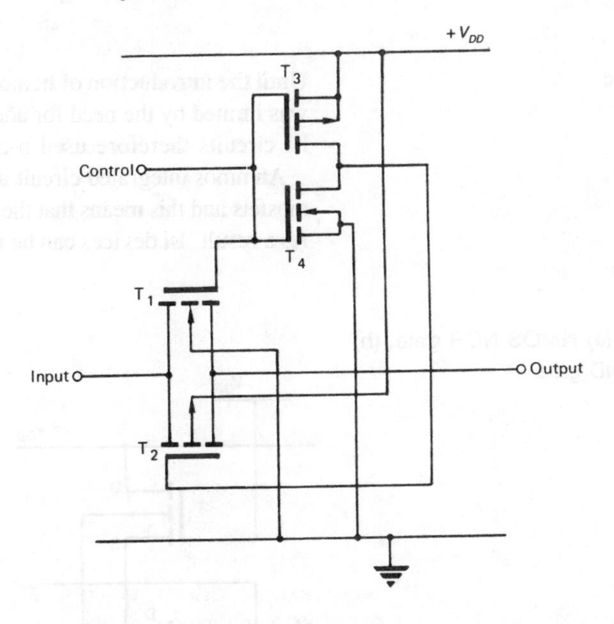

Cmos devices are available in both the A and B series, the B series operating from drain supply voltages of 3–20 V as opposed to 3–15 V for the A series. Also some of the electrical characteristics are slightly different. Some typical figures for cmos devices are

Low output voltage $V_{OL} = 0.05$ V max.
High output voltage $V_{OH} = V_{DD} - 0.05$ V min.
Noise margin 1 V min.
Propagation delay 30 ns

Sink current I_{OL} (V_{DD} = 5 V) = 1 mA
Sink current (V_{DD} = 15 V) = 6.8 mA
Source current (V_{DD} = 5 V) = 1 mA
Source current (V_{DD} = 15 V) = 6.8 mA

High-speed cmos

Since cmos was first introduced its fields of application have generally been restricted to circuits and systems in which the factors of over-riding importance are low power consumption and a high tolerance for power supply voltages.

A new version of cmos, known as **high-speed cmos** (hcmos), has recently been introduced on to the market. Hcmos combines the high speed of low-power Schottky ttl with the low power dissipation of standard cmos. The increased speed (by a factor of about 30) is obtained by the use of a new manufacturing technique. The technique also allows a much greater packing density than hitherto achieved in integrated circuitry and, consequently, hcmos is finding increased application in lsi and vlsi circuitry.

Nmos Logic

Until the introduction of hcmos, the packing density of cmos devices was limited by the need for adequate source-drain isolation and many lsi circuits therefore used n-channel mosfets only.

An nmos integrated circuit uses only n-channel enhancement-mode mosfets and this means that the fabrication process is relatively simple. As a result, lsi devices can be manufactured with a high packing den-

Fig. 2.26 (*a*) NMOS NOR gate, (*b*) NMOS NAND gate

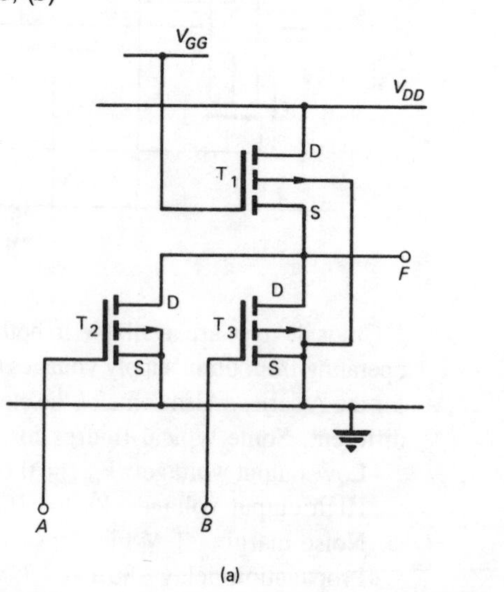

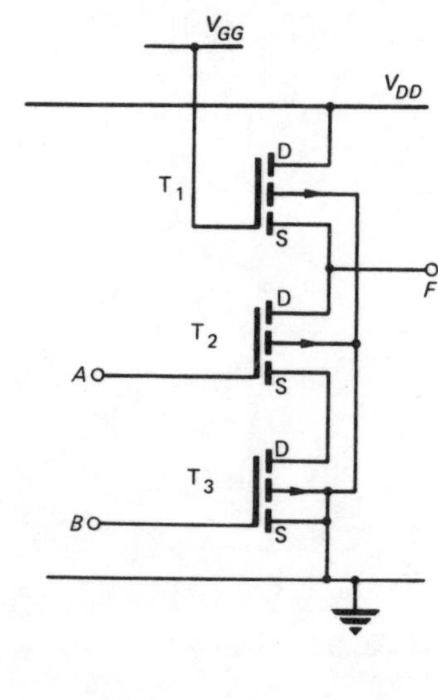

sity and consequent low cost. Figures 2.26(*a*) and (*b*) show the circuits of 2-input nmos NOR and NAND gates respectively. T_1 is biased, by the steady voltage $+V_{GG}$, to act as the *active load* resistor for the other transistors.

An n-channel enhancement-mode mosfet requires its gate to be held at a positive potential relative to its source for a drain current to flow. If the gate-source voltage is zero, the mosfet will be OFF.

Consider the *NOR gate* of Fig. 2.26(*a*); if both inputs are low then the associated transistors will not conduct current and the output voltage will be at the supply voltage V_{DD} volts.

When either or both inputs are high, both mosfets are ON and the output voltage is low. It is evident that the NOR function is performed.

The operation of the *NAND gate* (Fig. 2.26(*b*)) is very similar. When either or both inputs are low (0), the associated transistor does not conduct and the output voltage is high at $+V_{DD}$ volts or logical 1. Only when both of the inputs are high and both transistors are ON will the output voltage of the gate be at approximately 0 V. Thus the NAND function is performed.

Typically, an nmos gate will have a power dissipation of 50 μW and a propagation delay of 14 ns.

Nmos has been the major lsi technology for some time because of its high packing density and relative ease of design, and it has been widely employed for semiconductor memories and microprocessors. Often, an nmos circuit contains some depletion-mode mosfets as well as enhancement types in order to achieve a larger output voltage swing. However, as stated earlier, it now has a serious competitor in hcmos.

Tri-state Outputs

In some equipments, such as microprocessors, information is transferred from one point to another via buses. A number of different circuits may require to be connected to a bus but yet must not interact with one another. There is thus a need for circuits whose outputs can be directly connected in parallel with one another without affecting their operations.

Three-state (often called tri-state) circuits have *three* possible output states. Two of these are the usual logic 0 and logic 1 states, while the third state is one in which the output circuit is of high impedance. A *select* or *inhibit* terminal is provided to allow the output to be switched into, or out of, its high-impedance condition; see Fig. 2.27

Fig. 2.27 Tri-state inverter

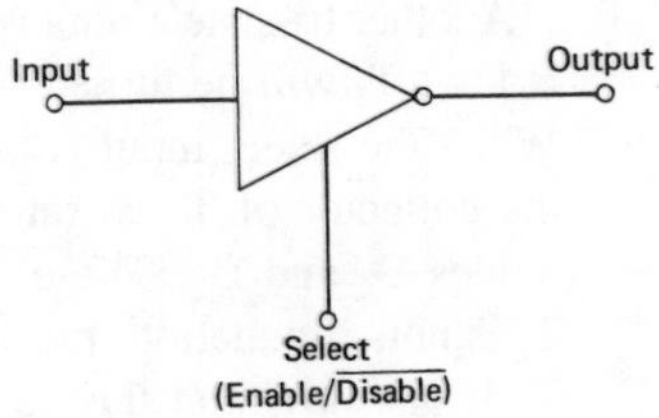

Input
Output
Select
(Enable/$\overline{\text{Disable}}$)

Table 2.2 Tri-state inverter

Input	*Select*	*Output*
0	1	High impedance
1	1	High impedance
0	0	1
1	0	0

which shows a tri-state inverter. The truth table for a tri-state inverter is given by Table 2.2.

Figure 2.28 shows a **ttl tri-state circuit**. When the select input is low (0), current flows through transistor T_1 to the low select input and there is then insufficient current passed to the base of T_2 to turn T_2 ON. With T_2 OFF, the base voltage of T_4 is low and hence T_4 is also turned OFF. The base potential of T_3 is held low by the select input and so T_3 is also non-conducting. Both the output transistors are in their non-conducting states and so the output impedance of the circuit is high.

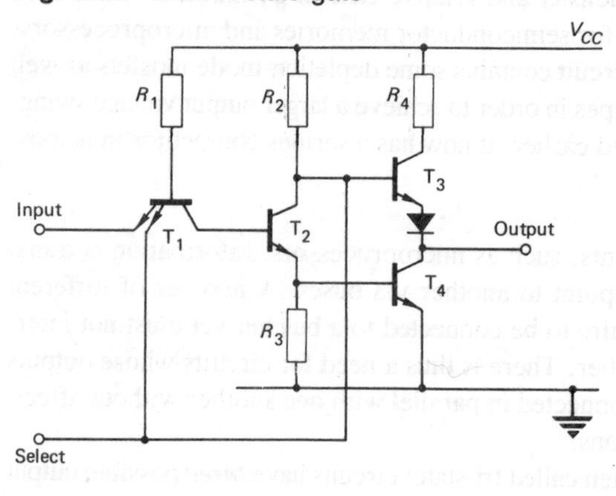

Fig. 2.28 TTL tri-state gate

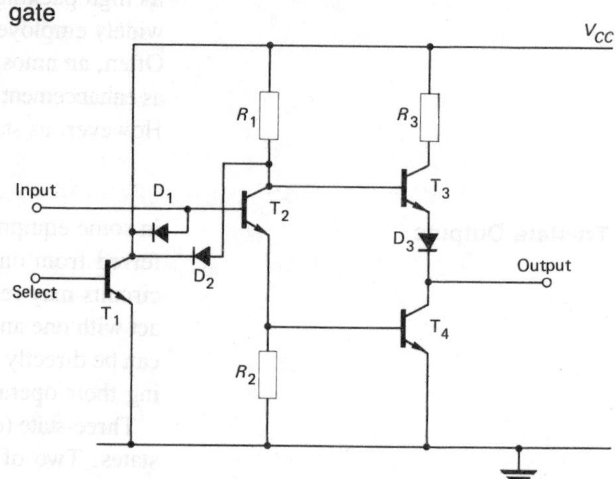

Fig. 2.29 Alternative ttl tri-state gate

If the select input of Fig. 2.28 is high, the circuit is able to operate in its normal manner and the output voltage is either high or it is low depending on the logic states of the inputs.

Another tri-state circuit is shown in Fig. 2.29. When the select input is low, T_1 will be turned OFF and the circuit will operate normally. When the select input is high, T_1 is turned ON. The low voltage at the collector of T_1 is transferred to the bases of T_2 and T_3 via the diodes D_1 and D_2. These two transistors therefore turn OFF. Since T_2 is non-conducting, the voltage across R_2 and hence at the base of T_4 is also low and this causes T_4 to also turn OFF.

Fig. 2.30 CMOS tri-state gate

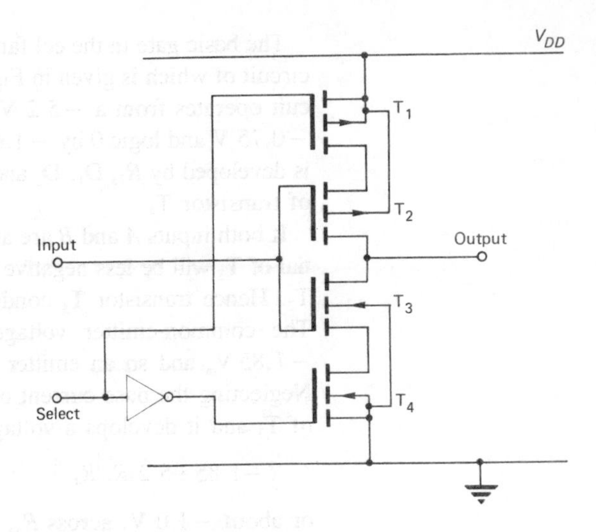

Figure 2.30 shows the circuit of a **cmos tri-state circuit**. When the select input is high, T_1 has its gate positive with respect to its source and turns OFF. T_4 has a gate-source voltage of zero and also turns OFF. The output impedance of the circuit is then very high. Conversely, when the select input is low, both T_1 and T_4 are turned ON and the operation of the circuit is determined by the signal applied to the input terminal.

Emitter-coupled Logic

The main feature of **emitter-coupled logic,** or **ecl**, is its very fast speed of operation with which only advanced Schottky ttl is a serious competitor. The ecl logic family generally finds application when the maximum possible speed of operation is the prime consideration. Very fast switching is obtained by ensuring that the transistors do not saturate when turned ON.

Fig. 2.31 ECL OR/NOR gate

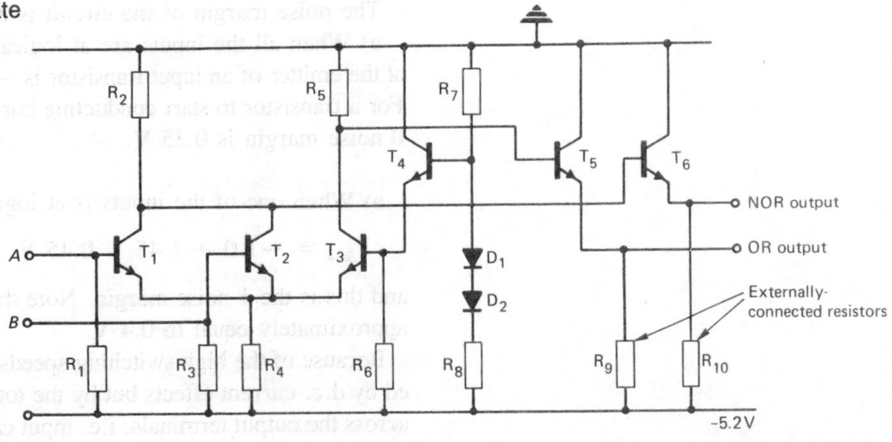

The basic gate in the ecl family is a combined **OR/NOR gate**, the circuit of which is given in Fig. 2.31. It should be noted that the circuit operates from a -5.2 V supply and logic 1 is represented by -0.75 V and logic 0 by -1.60 V. A reference voltage of -1.29 V is developed by R_7, D_1, D_2 and R_8 and is applied to the base terminal of transistor T_4.

If both inputs A and B are at logical 0 (-1.60 V), the base potential of T_3 will be less negative than the base potentials of both T_1 and T_2. Hence transistor T_3 conducts while T_1 and T_2 do not conduct. The common-emitter voltage is then equal to $-1.15-0.7 = -1.85$ V, and so an emitter current of $(-1.85 + 5.2)/R_4$ flows. Neglecting the base current of T_3, this is also the collector current of T_3 and it develops a voltage of

$$(-1.85+5.2)R_5/R_4$$

or about -1.0 V, across R_5, and this is sufficient to make T_5 conduct. T_5 is connected as an emitter follower and so its output voltage is

$$[(5.2 - 1.85)R_5/R_4] - V_{BE5} \text{ volts} \simeq -1.6 \text{ V or logical 0}$$

The collector potentials of T_1 and T_2 are at approximately 0 V and hence the output voltage of T_6 is

$$0 \text{ V} - V_{BE6} \simeq -0.75 \text{ V or logical 1}$$

If either input A or input B is at logical 1, i.e. -0.75 V, the base of the associated transistor is less negative than the base of T_3 and this transistor turns ON while T_3 turns OFF. The voltage dropped across R_5 is now zero and so the output voltage of T_5 switches to very nearly -0.7 V. The voltage developed across R_4 is

$$-0.75 - 0.7 = -1.45 \text{ V} \quad \text{and} \quad I_E \simeq I_C = (-1.45 + 5.2)/R_4$$

and so the voltage across R_2 is

$$+3.75R_2/R_4 \quad \text{and} \quad V_{out(T6)} = V_{R2} - V_{BE6}$$

The noise margin of the circuit is easily determined:

a) When all the inputs are at logical 0 or -1.60 V, the voltage at the emitter of an input transistor is -1.75 V so that V_{BE} is 0.15 V. For a transistor to start conducting current, $V_{BE} \simeq 0.5$ V so that the 0 noise margin is 0.35 V.

b) When one of the inputs is at logical 1 or -0.75 V, then

$$V_{BE4} = -1.0 + 1.45 = 0.45 \text{ V}$$

and this is the 1 noise margin. Note that the two noise margins are approximately equal to 0.4 V.

Because of the high switching speeds, the fan-out is not determined by d.c. current effects but by the total capacitance that is present across the output terminals, i.e. input capacitance of a gate times the number of gates. The fan-out is usually about 50 at the lower frequencies but falls as the desired speed of operation is increased.

Some standard ecl gates are the 1660 dual 4-input OR/NOR, the 1662 quad 2-input NOR, the 1664 quad 2-input OR, and the 1672 triple 2-input exclusive-OR. These ics have a power dissipation of 60 mW per gate and a propagation delay of 1.1 ns.

Wired-OR logic is often used with ecl stages.

A variation of ecl is known as *current mode logic* or *cml*. Cml uses voltage levels of 0V for logical 1 and −0.5 V for logical 0 and does not employ output emitter follower stages. Because of this, cml circuits are able to operate from a collector supply voltage of only −3.3 V. As a result cml is as fast to operate as ecl but it has a lower speed-power product.

Comparison between the Integrated Random Logic Families

The main characteristics of the integrated logic families used for random logic are listed in Table 2.3. Typical figures are quoted.

Table 2.3 Integrated logic families

Family	*Propagation delay (ns)*	*Power dissipation (mW)*	*Noise immunity (V)*	*Fan-in*	*Fan-out*	*Source current (mA)*	*Sink current (mA)*
Standard ttl	9	10	0.4	8	10	16	1.6
Schottky ttl	3	20	0.3	8	10	20	2.0
Schottky low-power ttl	7	2	0.3	8	10	8	0.4
cmos	30	0.001	1.5	8	50	1.6	0.001
ecl	1.1	40	0.4	5	25		
Advanced Schottky	1.5	2	0.3	8	10	20	0.5
Advanced low-power Schottky	4	1	0.4	8	10	8	0.1
hcmos	10	25×10^{-7}	depends on V_{DD}	8	50	4	0.001

Charge-coupled Devices

Figure 2.32(*a*) shows a p-type substrate on the top of which has been grown a narrow layer of silicon dioxide. Positioned on top of this insulating layer is a metal plate.

If a positive potential, greater than the threshold voltage V_t, is applied to the plate (Fig. 2.32(*b*)), a depletion layer or region will be formed in the substrate. The dotted line represents the *surface potential* which is equal to the voltage applied to the plate *minus* the

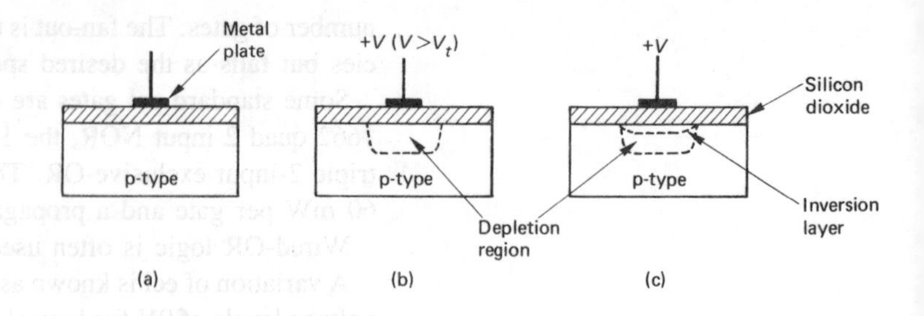

Fig. 2.32 Formation of an inverse layer

threshold voltage. Usually, $V_t \simeq 2$ V. Any free electrons in the depletion region, produced either by thermal agitation, or injected in a way to be described later, will be attracted to the area immediately beneath the plate. The electrons are then said to be stored within a *potential well* and form an *inversion layer* (Fig. 2.32(*c*)).

The **charge-coupled device** works by moving the charge stored in one potential well to another potential well, and this means that it must contain a large number, perhaps several hundreds, of metal plates or *gates*. Consider Fig. 2.33(*a*) which shows a p-type substrate with four metal plates mounted on to its silicon dioxide insulating layer. Three of the plates have a positive potential of +5 V applied to them but the second plate from the left has +10 V applied to it. All the available electrons, or charge, will be stored in the potential well beneath the second plate.

Fig. 2.33 Principle of a charge-coupled device

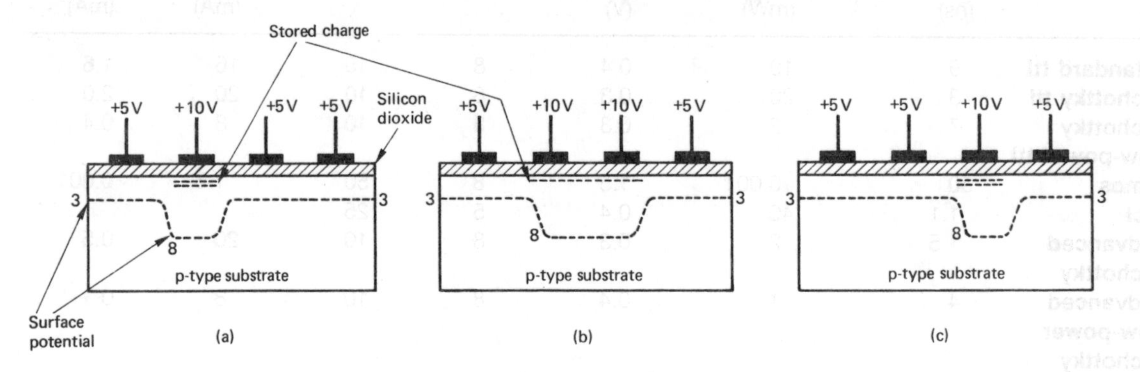

If the voltage applied to the third plate is increased to +10 V (Fig. 2.33(*b*)), the potential well beneath this plate will increase to the same depth as the well beneath the second plate. As a result the charge stored will be shared between the two potential wells as shown. Reducing the potential of the second plate to +5 V (Fig. 2.33(*c*)) means that the third plate is now at the most positive potential and so all the stored charge will be moved one place to the right into the potential well beneath the third plate. It should be clear that if the voltage on the fourth plate is now increased to +10 V, and then the potential of the third plate is reduced to +5 V, the stored charge can be moved one more place to the right.

Fig. 2.34 3-phase ccd system

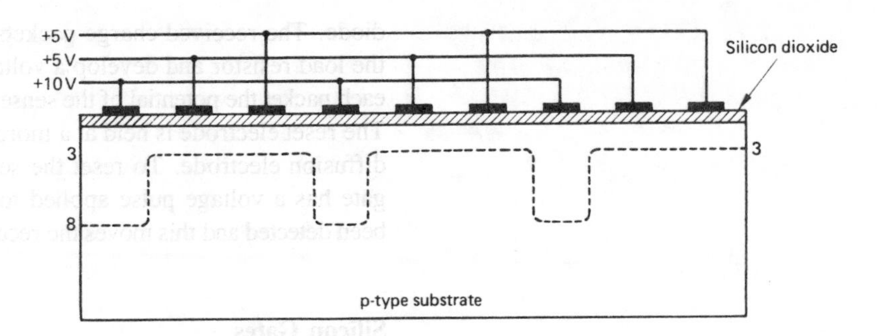

A practical charge-coupled device may have hundreds of gates and it is not possible to connect an individual positive potential to each one of them in turn. Instead, every third gate is commoned together (Fig. 2.34) to form a *three-phase system*. Clock waveforms of suitable phasing must be applied to each of the three lines in order to shift the charge along the device.

The method of injecting a packet of charge into the device varies; one method consists of applying a voltage pulse to the input gate that is of sufficiently high value to produce a momentary avalanche breakdown of the depletion region beneath the gate. For a binary 1 to be handled, the breakdown must occur at the instant that a potential well is present and is just about to move to the right. If no breakdown occurs as a well commences movement, then a binary 0 will be stored. Another method uses an input p-n junction; the p-type region is the substrate and the n-type region is a heavily-doped area into which electrons are injected by the application of a negative voltage (Fig. 2.35). Charge is placed into the potential well underneath the first transfer gate by applying a voltage greater than V_t to the input gate. The input diffusion region is normally held slightly reverse-biased and its function is to control the amount of charge placed into the well.

Fig. 2.35 Complete ccd system

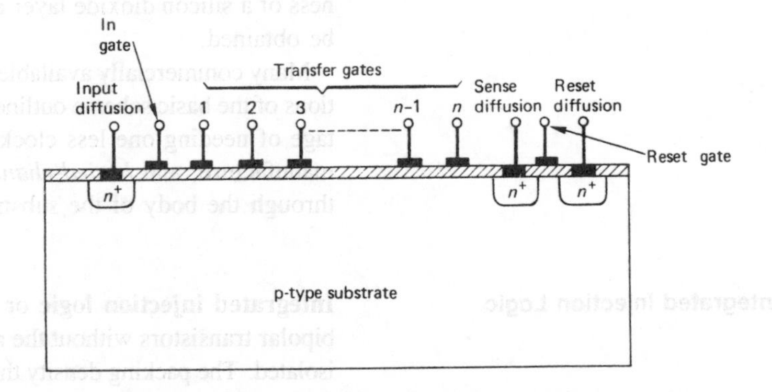

The arrival of charge packets at the output of the device is detected by connecting the sense diffusion electrode to a positive potential via a load resistor. The sense diffusion electrode acts as a reverse-biased diode. The received charge packets cause small currents to flow in the load resistor and develop a voltage across it. After the arrival of each packet the potential of the sense diffusion electrode must be reset. The reset electrode is held at a more positive potential than the sense diffusion electrode. To reset the sense diffusion electrode the reset gate has a voltage pulse applied to it after each charge packet has been detected and this moves the received charge to the reset electrode.

Silicon Gates

When a large number of gates are to be provided, and a typical ccd may well have several hundreds, difficulties arise with the fabrication of aluminium gates with *very* narrow gaps between them. The use of metallization has therefore been replaced, for most devices, by **silicon gates**.

Fig. 2.36 Silicon gate

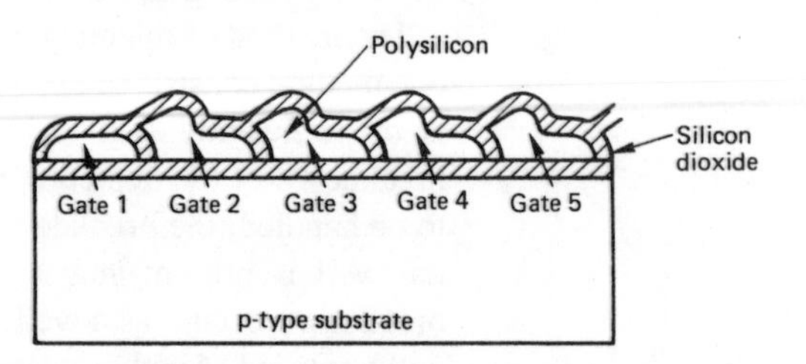

The principle of a silicon gate is shown by Fig. 2.36. A layer of polysilicon (an insulator) is deposited on top of the silicon dioxide layer and is then etched to form one third of the required number of gates, i.e. every third one. A layer of silicon oxide is formed and is then etched away as appropriate to form the remaining gates, i.e. 3, 6, 9, etc. and then, finally, a layer of silicon dioxide is grown overall. The separation between adjacent gates is now only the thickness of a silicon dioxide layer and is much less than could otherwise be obtained.

Many commercially available ccds employ one or both of two variations of the basic scheme outlined. *Two-phase* devices have the advantage of needing one less clock source but are more complicated to manufacture, and *buried-channel* devices move the charge packets through the body of the substrate rather than over its surface.

Integrated Injection Logic

Integrated injection logic or **I^2L** is a method of using integrated bipolar transistors without the need for each element to be separately isolated. The packing density that can be achieved is greater than even mosfet technology can achieve, and hence I^2L logic is eminently suitable for use in lsi devices. One of the main advantages of I^2L is that it is easily combined in the same chip, with any of the other logic families.

I^2L avoids the complexity of conventional bipolar transistor logic by the use of **inverted transistors**. This term simply means that the emitter and collector regions of a transistor are interchanged. A consequence of this is that I^2L transistors can have multiple collectors and so wired-OR logic is convenient.

Fig. 2.37 I^2L n-p-n transistor

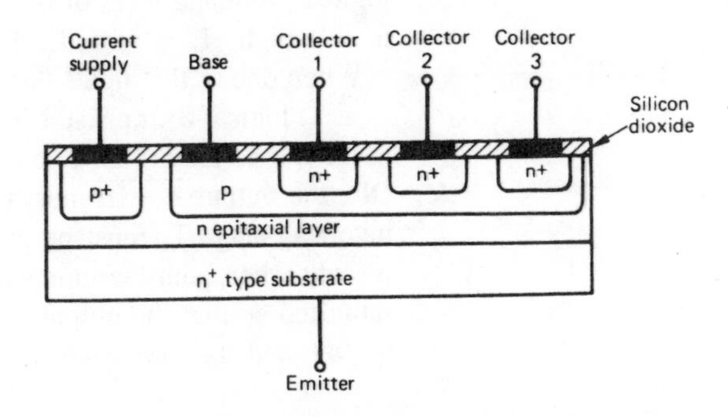

Fig. 2.37 shows the fabrication of an I^2L n-p-n transistor with three collectors. The emitter of the transistor is the n epitaxial layer; this layer extends throughout the chip and provides the emitter for *every* n-p-n transistor formed within the chip.

p-n-p transistors are used as current sources and these are of the *lateral* type [EIV]. The terminal marked as "current supply" is the emitter, the terminal marked "emitter" is the base, and the terminal labelled as "base" becomes the collector.

Fig. 2.38 I^2L OR/NOR gate

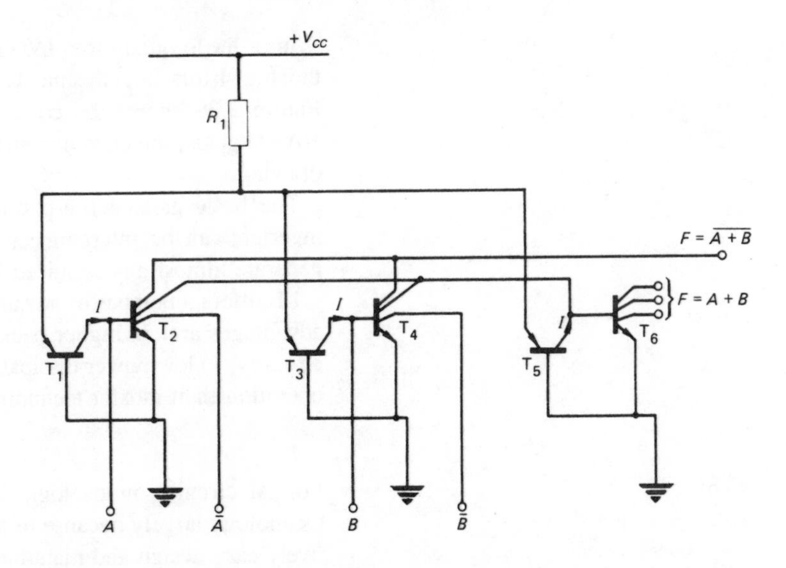

Fig. 2.38 shows a 2-input *OR/NOR gate* which also gives inversion of each input. The common base transistors T_1, T_3, and T_5 are operated as constant current sources that deliver a constant current to the base of each n-p-n transistor. The magnitude of the constant

current is determined by the value chosen for an externally connected resistor R_1. Only the one resistor is used to bias *all* the current sources within the ic and there could be several thousands of them.

Suppose both inputs are at logical 0 or approximately 0.05 V. Transistors T_2 and T_4 are then OFF and $\bar{A}$ and $\bar{B}$ outputs are high at the logical 1 voltage level of 0.75 V. The $F = \overline{A + B}$ output terminal is also high. T_6 is turned full ON and its output is at logical 0.

When one of the inputs, say A, is at logical 1 and the other input, B, is at logical 0, transistor T_1 will be saturated and T_4 will be OFF. Since the collectors of T_1 and T_4 are connected together in wired-OR, the output $\overline{A + B}$ must be at logical 0. Also the output of the inverting stage T_6 must be at logical 1.

Only when both the inputs are at logical 1 will both T_1 and T_4 be saturated so that the output of the circuit will be low. Now T_6 will be OFF and its output will be high.

Fig. 2.39 I²L AND/NAND gate

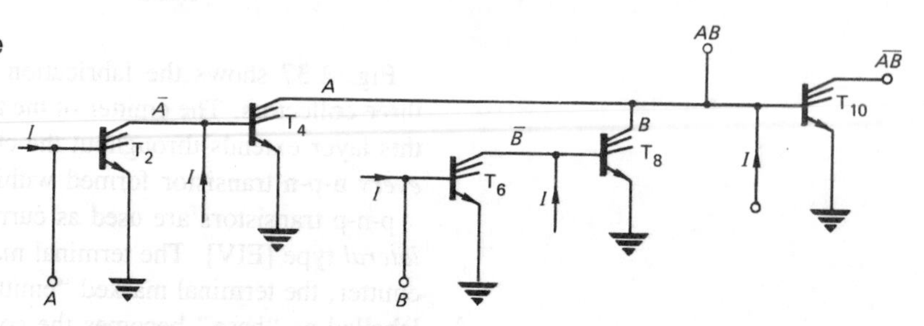

In order to obtain the *AND* and *NAND* functions using I²L logic, the transistors T_2, T_4 and T_6 must be interconnected differently. Figure 2.39 shows the connections needed to produce AND and NAND gates; the current sources have been omitted to simplify the drawing.

The basic gate, a p-n-p transistor current source and an n-p-n inverter, can be interconnected in a number of different ways to generate almost any required logical function.

I²L offers a number of advantages over other lsi techniques. These advantages are: *a*) higher packing density, *b*) compatibility with ttl circuitry, *c*) low power dissipation/operating speed product, *d*) faster operation than mosfet technology.

VLSI

For lsi circuits, nmos logic has generally been the predominant technology largely because of the high packing density and the relatively easy design and manufacture it allows. With the introduction of vlsi circuitry, involving the use of many thousands of mosfets, various drawbacks of nmos techniques have become apparent. Because of this, increasing interest in the use of cmos, instead of nmos, technology has developed.

For an nmos circuit to be able to produce an output voltage equal to the supply voltage, both depletion-mode and enhancement-mode mosfets must be used in the same chip. In contrast, a cmos circuit need employ only enhancement devices to be able to develop the same output voltage. Both nmos and cmos circuitry can be implemented using either dynamic or static logic but cmos circuits can often be designed to use fewer mosfets than their nmos equivalents. Power dissipation is one particular area in which cmos scores: a vlsi chip including some 20 000 transistors would dissipate about 1 W power using nmos static technology but only a few microwatts using cmos static logic. Because of this factor, nmos static logic is rarely employed for vlsi chips and dynamic logic is used instead. However, the use of dynamic logic introduces its own problems: there is then a need for periodic refreshing (p. 176) of the circuit and this means that extra circuit complexities are introduced and the effective packing density is reduced. On the other hand, there is a small speed advantage to be gained from the use of nmos techniques.

Exercises 2

2.1 *a*) Describe, with the aid of a circuit diagram, the operation of a ttl 3-input NAND gate.

b) Write down the truth table for the circuit given and state how the logical function of the circuit changes if negative logic is used.

c) How does ttl compare with cmos in respect of (i) noise immunity, (ii) switching speed, (iii) power consumption, (iv) fan-out?

2.2 Describe, with the aid of a circuit diagram, the operation of a 3-input cmos NAND gate.

Draw up a table to show how the speed, power dissipation and other parameters of Schottky low-power ttl compare with cmos logic.

2.3 With the aid of a circuit diagram and a truth table explain the operation of an ecl OR/NOR gate.

In the circuit of Fig. 2.31, $R_2 = 290\Omega$, $R_4 = 1180\Omega$, $R_5 = 300\Omega$. If the reference voltage at the base of T_4 is -1.29 V, calculate the output voltage of the circuit *a*) when it is high, *b*) when it is low. Take logic 1 as -0.75 V and logic 0 as -1.60 V.

Fig. 2.40

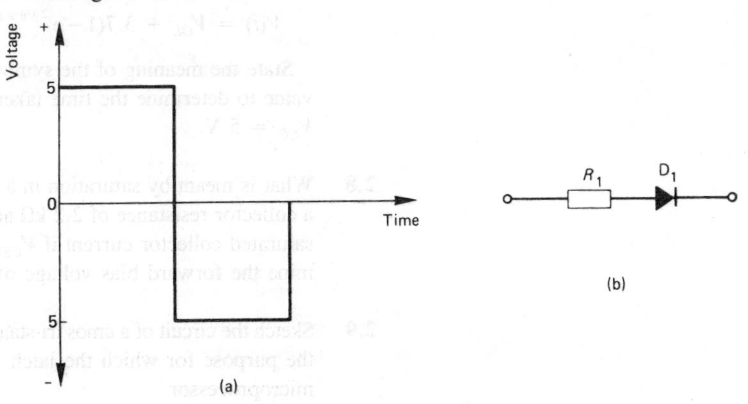

2.4 Explain with the aid of static characteristics how a diode can be used as a switch.

The voltage pulse shown in Fig. 2.40(*a*) is applied to the circuit of Fig. 2.40(*b*). Explain, with the aid of suitable diagrams, the voltage that appears across the diode.

2.5 Explain the purpose of the totem pole output stage used in a ttl gate. The transistor used in the circuit of Fig. 2.41 has the following parameters: $V_{CE(SAT)} = 0.2$ V and $V_{BE(SAT)} = 0.6$ V. Calculate *a*) the current flowing in the transistor T_3 when the output is low if the diode D_1 were not present, *b*) the minimum current gain for T_3 for the gate to be able to change state.

Fig. 2.41

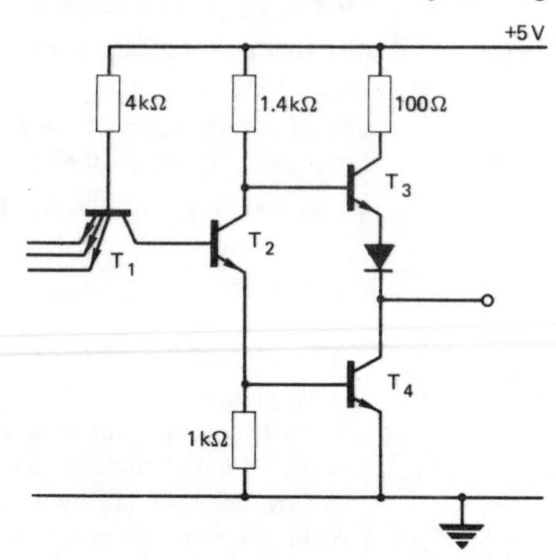

2.6 A switching transistor has the following data:

$V_{CB(max)} = 20$ V, $V_{CE(max)} = 20$ V, $I_{C(max)} = 450$ mA, $P_{Diss(max)} = 520$ mW
$V_{CE(SAT)} = 250$ mV, $V_{BE(SAT)} = 800$ mV
Rise time = 22 ns Storage time = 100 ns
Delay time = 6 ns Fall time = 28 ns

Draw some typical characteristics and hence show how the given ratings limit the choice of supply voltage and collector load resistance.

Explain the meaning of each of the four switching times mentioned.

2.7 The rising output voltage of a low-power Schottky ttl gate is given by

$$V(t) = V_{OL} + 3.7(1 - e^{-t/8 \times 10^{-9}})$$

State the meaning of the symbol V_{OL} and quote a typical value. Use this value to determine the time taken for the output voltage to rise to 90% of $V_{CC} = 5$ V.

2.8 What is meant by saturation in a bipolar transistor circuit? A transistor has a collector resistance of 2.2 kΩ and a supply voltage of 12 V. Calculate the saturated collector current if $V_{CE(SAT)} = 0.2$ V. If $V_{BE(SAT)} = 0.75$ V determine the forward bias voltage of the collector/base junction.

2.9 Sketch the circuit of a cmos tri-state latch and explain how it functions. Indicate the purpose for which the latch is used on the bus lines employed with a microprocessor.

3 Combinational Logic Circuits

Introduction

Many digital circuits can be constructed using a number of suitably interconnected logic elements such as gates and bistable multivibrators (see Chapter 4). If a circuit does not involve a need for time delay or storage, only gates of one kind are necessary. Many of the more common digital circuits, such as binary adders, multiplexers and decoders for example, are packaged as integrated circuit devices in the various logic families.

In this chapter the various ways in which gates can be interconnected to produce more complex digital functions will be described and, where appropriate, details of corresponding msi or lsi circuits will be given.

The design process for a digital circuit starts with the writing down of the truth table of the required logical operation. Each 1 that appears in the output column of the truth table must be represented by a term in the Boolean equation that describes the operation of the circuit. The Boolean equation must contain each literal that is at 1 and the complement of each literal that is at 0.

Binary Arithmetic Circuits

Very often in digital circuitry the need arises for the addition, subtraction, multiplication or division of two binary numbers. For all of these arithmetic processes the basic circuit employed is the *binary adder* since it can be modified, with additional logic circuits, to perform the other functions.

Binary Addition

Half-adder

The simplest form of binary adder is known as the **half-adder**. This has two inputs A and B but it does not have a "carry input" from some previous stage. The half-adder can be used to add together the two least significant bits A_0 and B_0 of two numbers where there is no input carry. The block diagram of a half-adder is shown in Fig. 3.1.

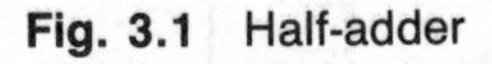

Fig. 3.1 Half-adder

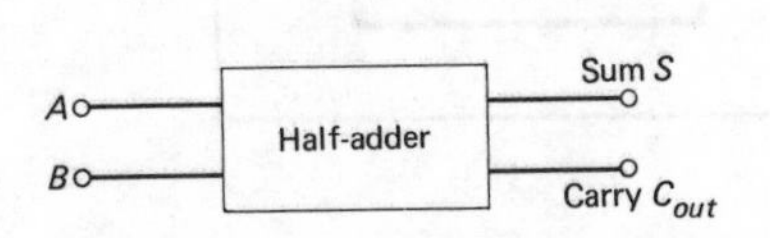

The truth table of a half-adder is given by Table 3.1. Two different Boolean equations can be obtained from Table 3.1 that express the sum and the carry. These equations are

$$S = A\bar{B} + \bar{A}B \qquad C_{out} = AB \tag{3.1}$$

Table 3.1 Half-adder truth table

A	0	1	0	1
B	0	0	1	1
Sum	0	1	1	0
Carry C_{out}	0	0	0	1

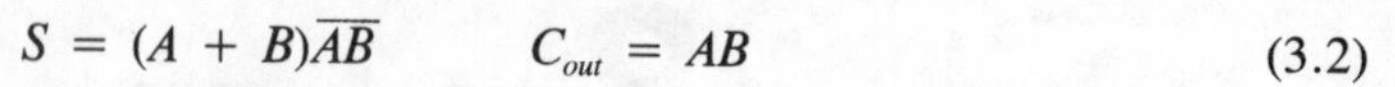

$$S = (A + B)\overline{AB} \qquad C_{out} = AB \tag{3.2}$$

These equations can be directly implemented using a combination of AND and OR gates and inverters, but for the reasons given in Chapter 1 it is more likely that either NAND or NOR gates would be exclusively employed. Since equation (3.1) is in the sum-of-products form it is best suited to a NAND gate implementation; while equation (3.2) is easier to implement using NOR gates.

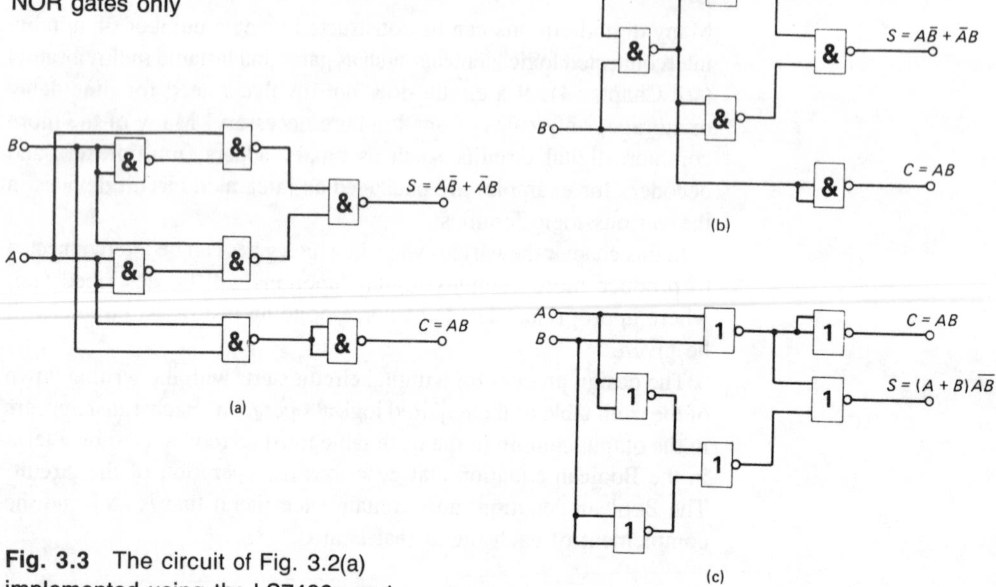

Fig. 3.2 Half-adder circuits (*a*) and (*b*) using NAND gates only, (*c*) using NOR gates only

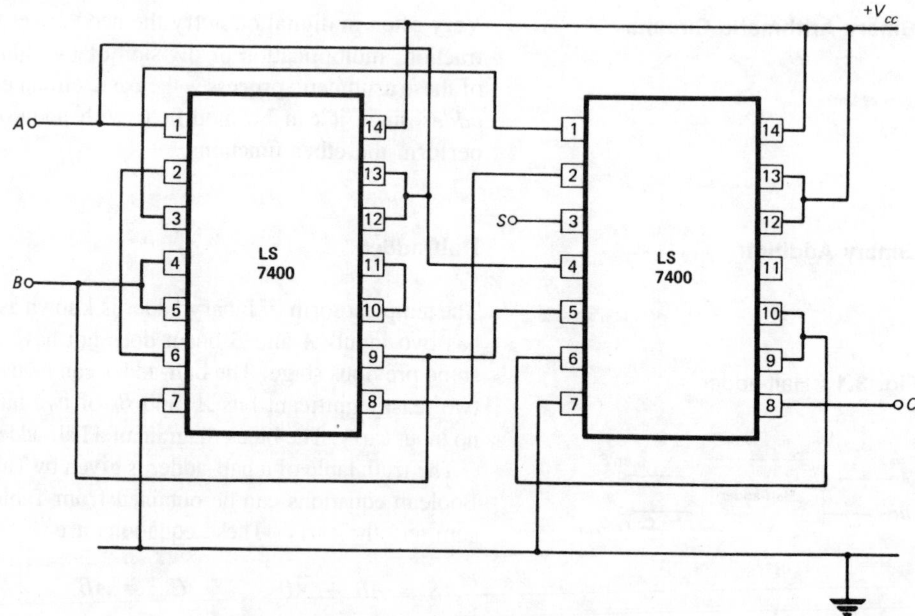

Fig. 3.3 The circuit of Fig. 3.2(a) implemented using the LS7400 quad 2-input NAND gate

Using the rules enunciated on page 23, Figs. 3.2(*a*) and (*c*) can be drawn directly. A simpler alternative to Fig. 3.2(*a*) is given in (*b*).

As an example of how these circuits could be fabricated using ttl or cmos devices, Fig. 3.3 shows Fig. 3.2(*a*) implemented using the low-power Schottky ttl integrated circuit LS 7400 quad 2-input NAND gate.

Full-adder

Fig. 3.4 Full-adder

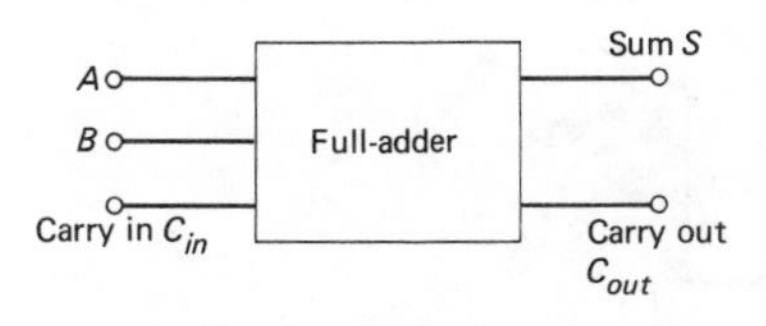

The full-adder is a circuit which adds two input bits A and B together with a possible carry input bit from a previous stage. The block diagram of a full-adder is shown in Fig. 3.4, and its truth table is given by Table 3.2. From it, Boolean expressions for the sum S and the output carry C_{out} can be obtained:

$$S = A\bar{B}\bar{C}_{in} + \bar{A}B\bar{C}_{in} + \bar{A}\bar{B}C_{in} + ABC_{in}$$

$$= (A\bar{B} + \bar{A}B)\bar{C}_{in} + (\bar{A}\bar{B} + AB)C_{in} \quad (3.3)$$

$$= (\text{exclusive-OR})\ \bar{C}_{in} + (\text{exclusive-NOR})\ C_{in} \quad (3.4)$$

Table 3.2 Full-adder truth table

A	0	1	0	0	1	1	0	1
B	0	0	1	0	1	0	1	1
C_{in}	0	0	0	1	0	1	1	1
Sum	0	1	1	1	0	0	0	1
C_{out}	0	0	0	0	1	1	1	1

Also,

$$C_{out} = AB\bar{C}_{in} + A\bar{B}C_{in} + \bar{A}BC_{in} + ABC_{in}$$

$$= AB(\bar{C}_{in} + C_{in}) + C_{in}(A\bar{B} + \bar{A}B)$$

$$= AB + C_{in}(\text{exclusive-OR}) \quad (3.5)$$

The expressions obtained for the sum and the carry-out can be implemented using NAND or NOR gates only, or use can be made of exclusive-OR gates. The NAND gate implementation of the full-adder is shown in Fig. 3.5.

Equation (3.4) can be implemented, somewhat more simply, by using exclusive-OR gates, e.g. the 7486 quad 2-input, connected as shown by Fig. 3.6. It is not obvious that this circuit achieves the desired result but the output of the right-hand exclusive-OR gate is

$$S = (A\bar{B} + \bar{A}B)\bar{C}_{in} + C_{in}(AB + \bar{A}\bar{B})$$

and

$$C_{out} = \overline{\overline{AB}\ \overline{C_{in}(A\bar{B} + \bar{A}B)}}$$

$$= AB + C_{in}(A\bar{B} + \bar{A}B)$$

A full-adder can also be constructed by connecting two half-adders together in the manner shown by Fig. 3.7. The two outputs of the left-hand half-adder are

$$\text{Sum } A = A\bar{B} + \bar{A}B \quad \text{and} \quad \text{Carry } A = AB$$

Hence, for the right-hand half-adder,

$$\text{Sum } B = (A\bar{B} + \bar{A}B)\bar{C}_{in} + \overline{(A\bar{B} + \bar{A}B)}C_{in}$$

Fig. 3.5 Full-adder using NAND gates only

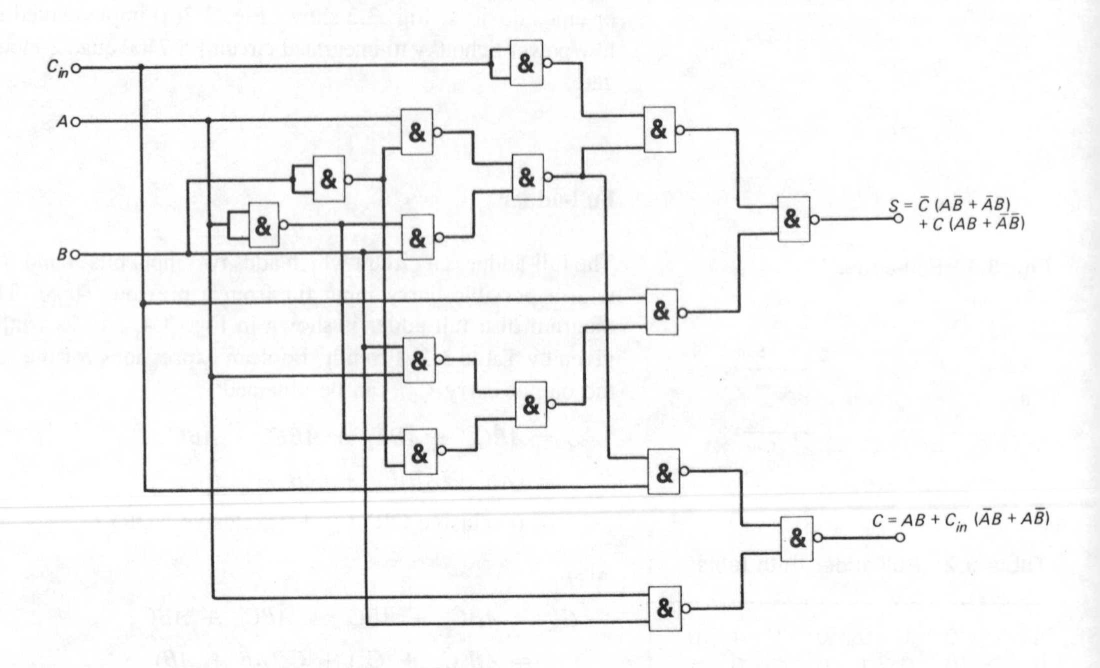

Fig. 3.6 Full-adder using exclusive-OR and NAND gates

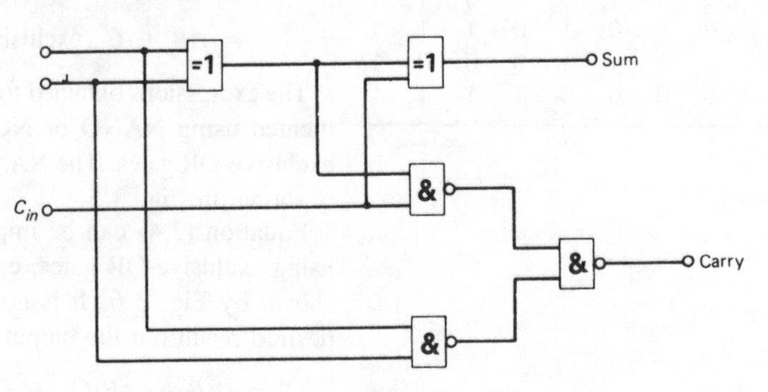

3.7 Connection of two half-adders to produce one full-adder

$$= (A\bar{B} + \bar{A}B)\bar{C}_{in} + (AB + \bar{A}\bar{B})C_{in}$$

which is equation (3.4) and the sum output of the circuit. Also,

$$\text{Carry } B = (A\bar{B} + \bar{A}B)C_{in}$$

This means that the carry-out of the circuit is

$$C_{out} = AB + (A\bar{B} + \bar{A}B)C_{in}$$

which is equation (3.5).

Complete full-adder circuits are also available in the ttl and cmos families. The ttl 7482 2-bit adder consists of three inverters, fourteen AND gates, and four NOR gates, showing clearly the advantages of using an msi device.

Adding Multi-bit Binary Numbers

When two multi-bit binary numbers are to be added, two different techniques are available. These two methods are known as *parallel addition* and *serial addition*. Parallel addition requires the use of *n* full-adders to add two *n*-bit binary numbers and it is very fast in its operation. Serial addition, on the other hand, uses only one full-adder but needs some short-term memory, which is provided by *shift registers* (p. 143).

Parallel Adder

The way in which 4 full-adders should be connected to form a 4-bit **parallel adder** is shown by Fig. 3.8.

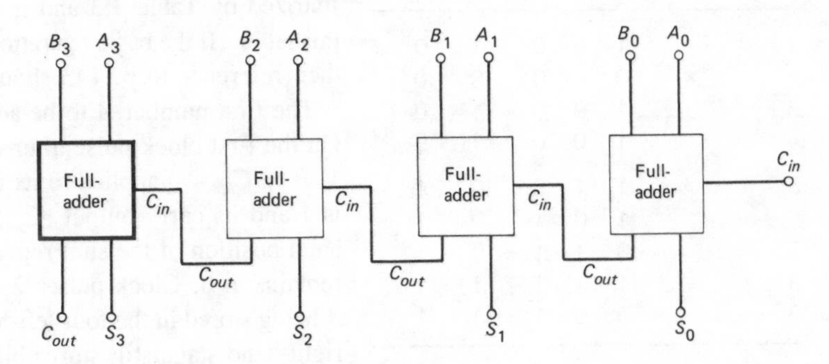

Fig. 3.8 4-bit parallel full-adder

Two 4-bit binary numbers $A_3A_2A_1A_0$ and $B_3B_2B_1B_0$ are applied to the *A* and *B* inputs of the four full-adders. A_0 and B_0 are, of course, the least significant bits. Any carry from a previous stage is applied to the C_{in} terminal of the right-hand (least significant) full-adder. The carry-out C_{out} terminal of each full-adder is directly connected to the C_{in} terminal of the next more significant full-adder. The C_{out} terminal of the left-hand (most significant) full-adder provides the carry-out of the complete circuit. The carry-out bits pass through the circuit stage by stage and an individual sum will only be correct when the carry (if any) from the preceding stage has been applied. This means that the output sum and carry of the circuit must not be read until enough time has elapsed, from the instance the input numbers were applied, for all of the carry bits to have propagated through the circuit.

Serial Adder

The block diagram of a 4-bit **serial adder** is shown in Fig. 3.9. The two 4-bit numbers to be added are applied serially to the input register of the circuit, and after nine clock pulses have elapsed the required sum is stored in the sum register.

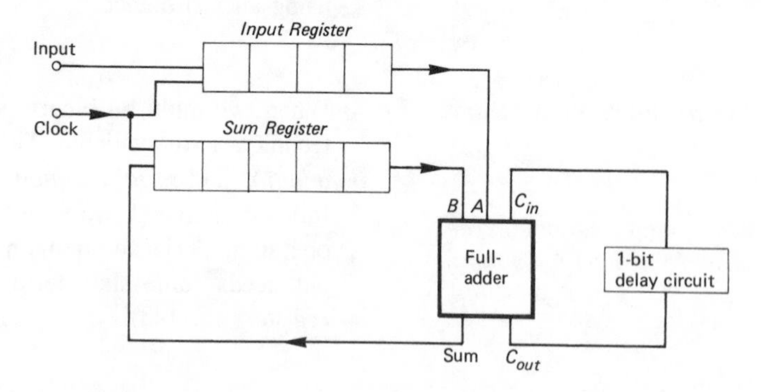

Fig. 3.9 4-bit serial full-adder

Table 3.3

Clock pulse	A	B	C_{in}	Sum	C_{out}
1	1	0	0	1	0
2	0	0	0	0	0
3	1	0	0	1	0
4	1	0	0	1	0
5	1	1	0	0	1
6	1	0	1	0	1
7	0	1	1	0	1
8	1	1	1	1	1
9	0	0	1	1	1

The operation is perhaps best described by considering the addition of two particular binary numbers. Suppose the two numbers are $C = 1101$ and $D = 1011$. The sequence of the operation is summarized by Table 3.3 and it will be described in conjunction with this table. (If the basic operation of the shift register is not understood then reference to p. 143 should first be made.)

The first number A to be added is loaded into the input register. On the first clock pulse after this (pulse 1), the full-adder has $A = 1$, $B = C_{in} = 0$ applied to its input terminals, so that its sum output is 1 and its carry output (C_{out}) is 0. The 1 bit is stored in the left-hand position of the sum register; thus the B input to the full-adder remains at 0. Clock pulses 2, 3 and 4 result in the binary number A being stored in the four left-hand positions of the sum register. The right-hand stage still stores binary 0 so that the B input to the full-adder remains at 0. The second binary number D is now held in the input register.

When the fifth clock pulse arrives, the data held in the sum register shifts one place to the right and then both inputs to the full-adder are at 1. Since $C_{in} = 0$, the sum S is 0 and hence C_{out} is 1. This carry-out value is only applied to the C_{in} input after a time delay equal to the period occupied by one clock pulse. Hence, after the sixth clock pulse has occurred, $A = C_{in} = 1$ and $B = 0$; then the sum $S = 0$ and $C_{out} = 1$. On the seventh clock pulse, $A = 0$ and $B = C_{in} = 1$, so once again the sum $S = 0$ and $C_{out} = 1$. After the ninth clock pulse, the sum register holds the sum of the two input numbers, i.e. 11000. Notice that, although the two input numbers each have four bits, their sum has five bits and this is why the sum register has an extra stage.

An alternative form of serial adder is shown in Fig. 3.10. The opera-

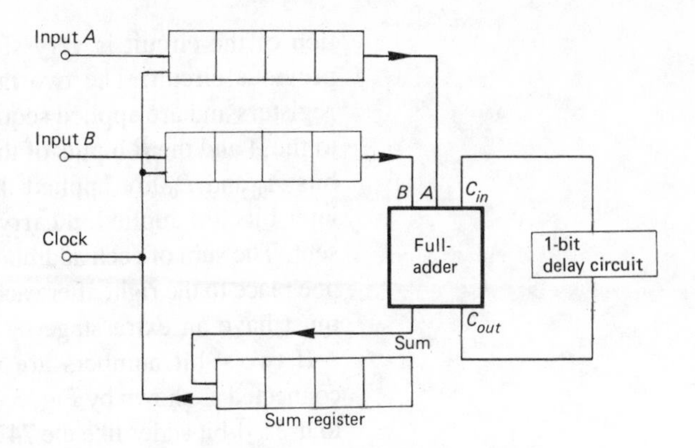

Fig. 3.10 Alternative 4-bit serial full-adder

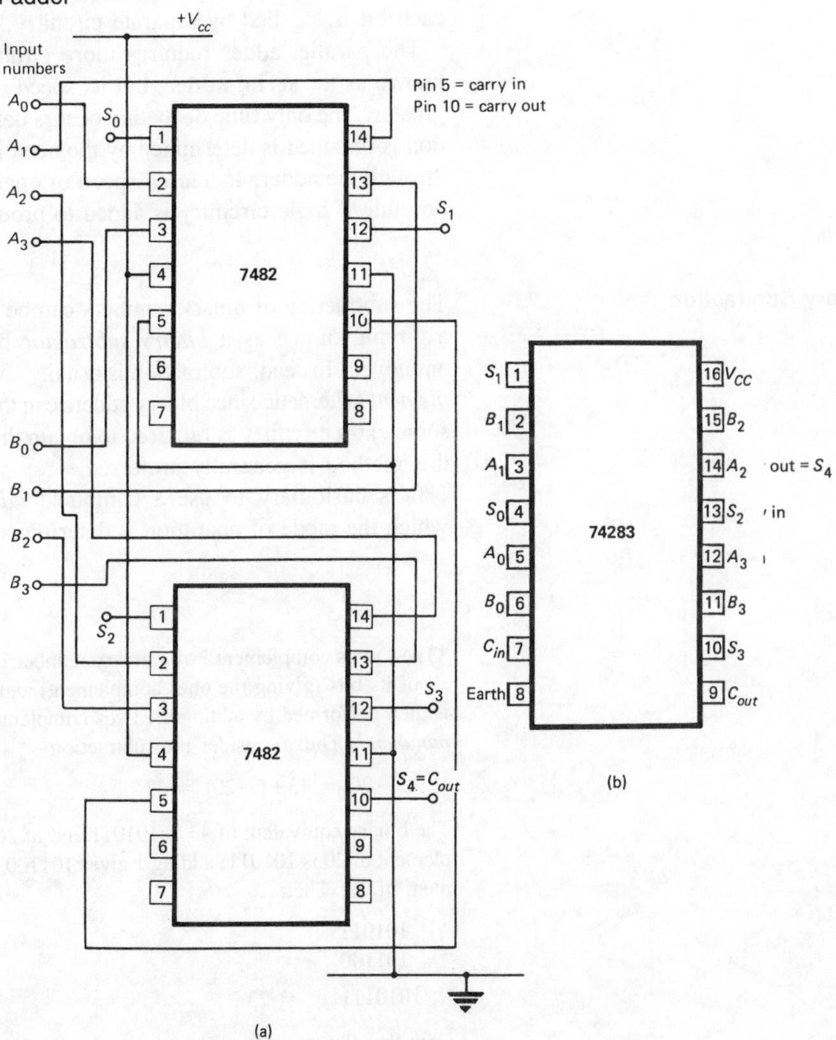

Fig. 3.11 4-bit full adder circuits (*a*) using two 7482 2-bit full-adders, (*b*) using one 74283 4-bit full-adder

tion of the circuit is very similar to that already described for the previous circuit. The two numbers A and B are held in the input registers and are applied sequentially, under the control of the clock, to the A and the B inputs of the full-adder. When the least significant bits A_0 and B_0 are applied there will be no carry input C_{in}. When later bits are applied, a carry from a previous addition *may* be present. The sum of each addition is stored in the sum register and shifted one place to the right after each clock pulse. As before the sum register must have an extra stage.

If two 4-bit numbers are to be added, two 7482s can be interconnected as shown by Fig. 3.11(*a*), but it would be easier and cheaper to use a 4-bit adder like the 74283 (Fig. 3.11(*b*)). Note that the carry-in pin is connected to earth when there is no input carry. In both circuits A_0 and B_0 are the least significant bits.

This method of binary addition is known as *parallel addition* since each bit is handled by separate circuitry.

The parallel adder requires more circuitry than the alternative, known as the **serial adder**, but its speed of operation is very much greater. The only time delay that occurs before the result of the addition is obtained is determined by the need for the carry to propagate through the adder. Increased speed of operation at increased cost is possible if logic circuitry is added to produce "carry look-ahead".

Binary Subtraction

The subtraction of binary numbers can be performed directly using a circuit known as a *binary subtractor* but this method is rarely employed. Instead, subtraction is usually carried out using *twos complement* arithmetic since binary adders can then be used.* Even though some extra circuitry is required to obtain the necessary complements this method is generally preferred.

It is customary to use a combined *adder/subtractor circuit* in which the mode of operation is determined by a control signal (see

*The "twos complement" of a binary number is obtained by first inverting all of its bits (giving the ones complement) and then adding 1. Subtraction is then performed by adding the twos complement of the *subtrahend* to the *minuhend*. Thus, consider the subtraction

$$43-20 = 43+(-20) = 23$$

The binary equivalent of 43 is 101011 and of 20 is 010100. The ones complement of 20 is 101011; adding 1 gives 101100 and this is the twos complement of 20. Then

$$\begin{array}{r} 101011 \\ 101100\ + \\ \hline 1010111 \end{array} = 23$$

Note that the most significant bit is dropped.

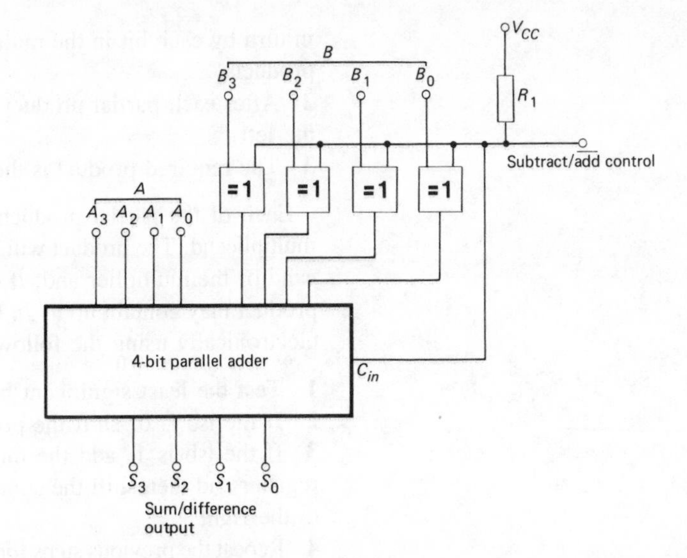

Fig. 3.12 Binary adder/subtractor circuit

Fig. 3.12). This figure shows how a full-adder is connected to act as either an adder or a subtractor. When the control signal is low, the carry-in C_{in} terminal will be low and the exclusive-OR gates will act to pass each input through to the full-adder itself. The inputs are not altered in any way and the circuit acts as an adder. When the control line is high, the C_{in} terminal will also be high and each of the exclusive-OR gates will invert its input signal. Thus the B input number is inverted *and* has 1 added to it and so it is changed into its twos complement form.

A complete adder/subtractor circuit can be obtained in a single ic package, e.g. the 74385.

Binary Multiplication

When two binary numbers are to be multiplied together, the method commonly employed is one of "adding and shifting". Consider, as an example, the determination of the product 1101 × 1001. The first number, 1101, is known as the multiplicand and the other number, 1001, is called the multiplier. The procedure used to perform this calculation is

```
      1101
      1001  ×
      1101
     0000
    0000
   1101
   1110101  (Ans)
```

The steps are as follows:

1 Multiply the multiplicand by first the least significant bit and then

in turn by each bit in the multiplier. This gives a number of partial products.

2 After each partial product has been obtained, shift one place to the left.

3 The required product is then the sum of all the partial products.

Each of the partial products is *either* equal to 0 *or* equal to the multiplicand. The product will have more bits than either the multiplicand or the multiplier and, if each of these numbers has n bits, the product may contain up to $2n$ bits. The process can be implemented electronically using the following steps:

1 Test the least significant bit of the multiplier.

2 If the lsb is 0, shift the product register one place to the right.

3 If the lsb is 1, add the multiplicand to the msb of the product register and then shift the contents of the product register one place to the right.

4 Repeat the previous steps for each of the bits stored in the multiplier register.

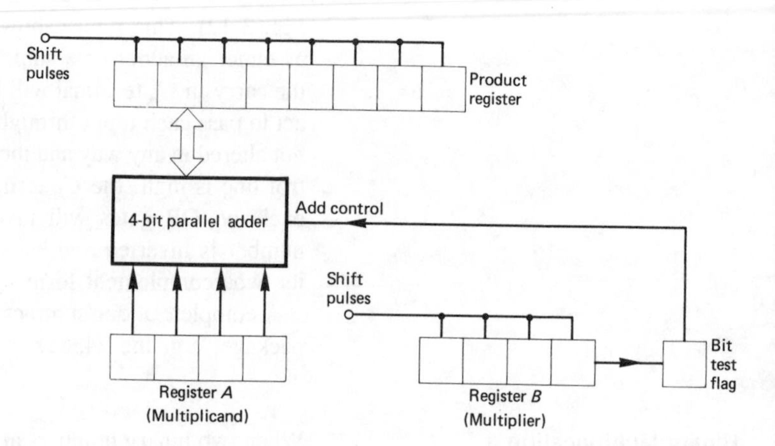

Fig. 3.13 Binary multiplier

The circuit of a binary multiplier based on this principle is shown in Fig. 3.13. The multiplicand is first loaded into the register A, the multiplier is loaded into register B, and the product register is cleared.

1 The lsb held in register B is then tested. If it is flagged as a 1, the add control instructs the binary adder to add the contents of register A to the contents (0000) of the product register (PR). If it is flagged as a 0, the contents of the product register are unchanged. In this case the results are

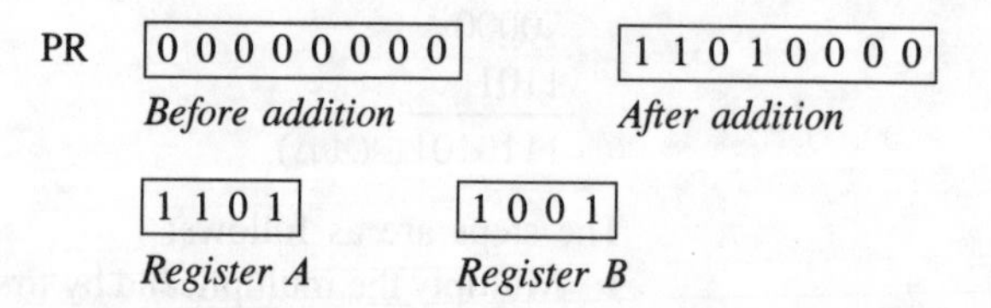

The contents of both the product register and the register *B* are now shifted one place to the right (with the result that the lsb B_0 is lost).

2 The right-hand bit held in register *B* is now tested to determine whether it is 1 or 0. In this case it is 0 so that the contents of the product register are left unchanged. The bits held in each register are now:

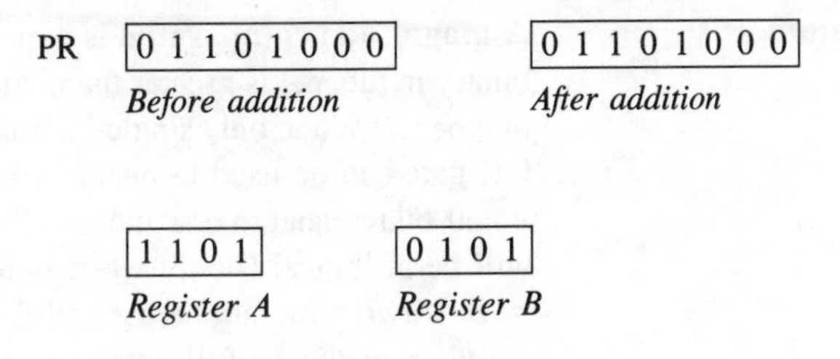

The contents of both the product and the *B* registers are now shifted one place to the right.

3 The steps listed in **1** and **2** are repeated until all four bits of the multiplier have been tested. The results for the particular example in question are:

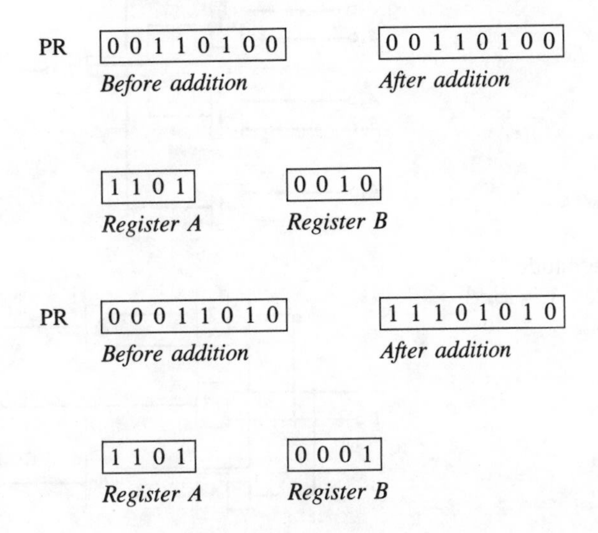

When the contents of the product and *B* registers once more move one place to the right, their contents are:

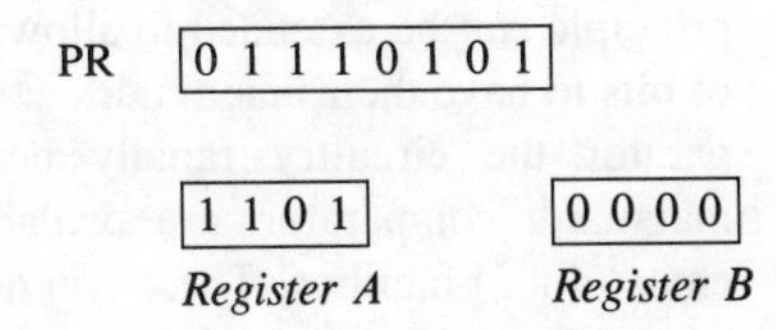

It will be seen that the product register now holds the correct product.

Several ic binary multipliers are available, two examples being the 74784 8-bit serial/parallel multiplier and the 74384 twos complement serial/parallel multiplier.

Magnitude Comparators

A **magnitude comparator** is a circuit that determines whether one binary number A is greater than, equal to, or less than another binary number B. When only single-bit numbers are involved, the exclusive-OR gate can be used to decide whether or not two binary numbers A and B are equal to one another. The output of an exclusive-OR gate will be at logical 0 only when both of its input are equal, i.e. if $A = B$. If two n-bit numbers A and B are to have their magnitudes compared, n exclusive-OR gates can be used whose outputs fan into a single NOR gate to give a logical 1 output only when $A = B$ (see Fig. 3.14 which shows the circuit for the case when $n = 4$).

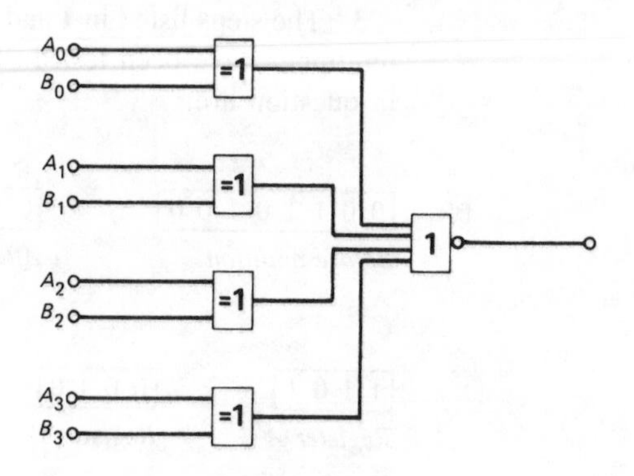

Fig. 3.14 Magnitude comparator using exclusive-OR gates (indicates $A = B$ or $A \neq B$ only)

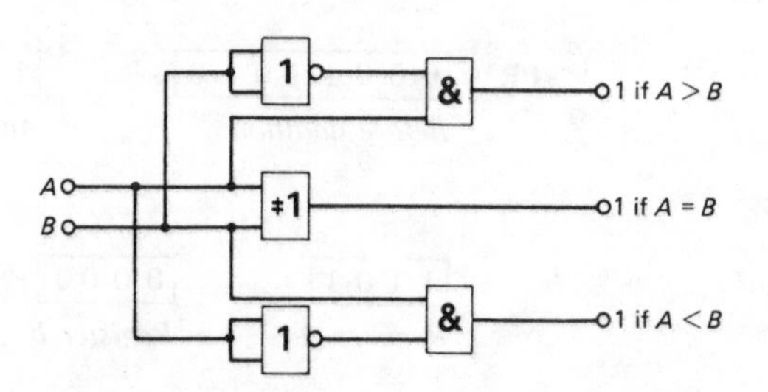

Fig. 3.15 1-bit magnitude comparator (indicates $A = B$, $A > B$, or $A < B$)

If, when A and B are not equal, some indication is required as to whether $A > B$ or $A < B$, then extra circuitry is necessary. Figure 3.15 shows one possible arrangement for a 1-bit comparator. The same principle *can* be extended to allow numbers containing any number of bits to have their magnitudes compared but even for a 4-bit comparator the circuitry rapidly becomes very complicated. 4-bit magnitude comparators are available in both the ttl (7485) and the cmos (4063) families. These two devices have identical pin connec-

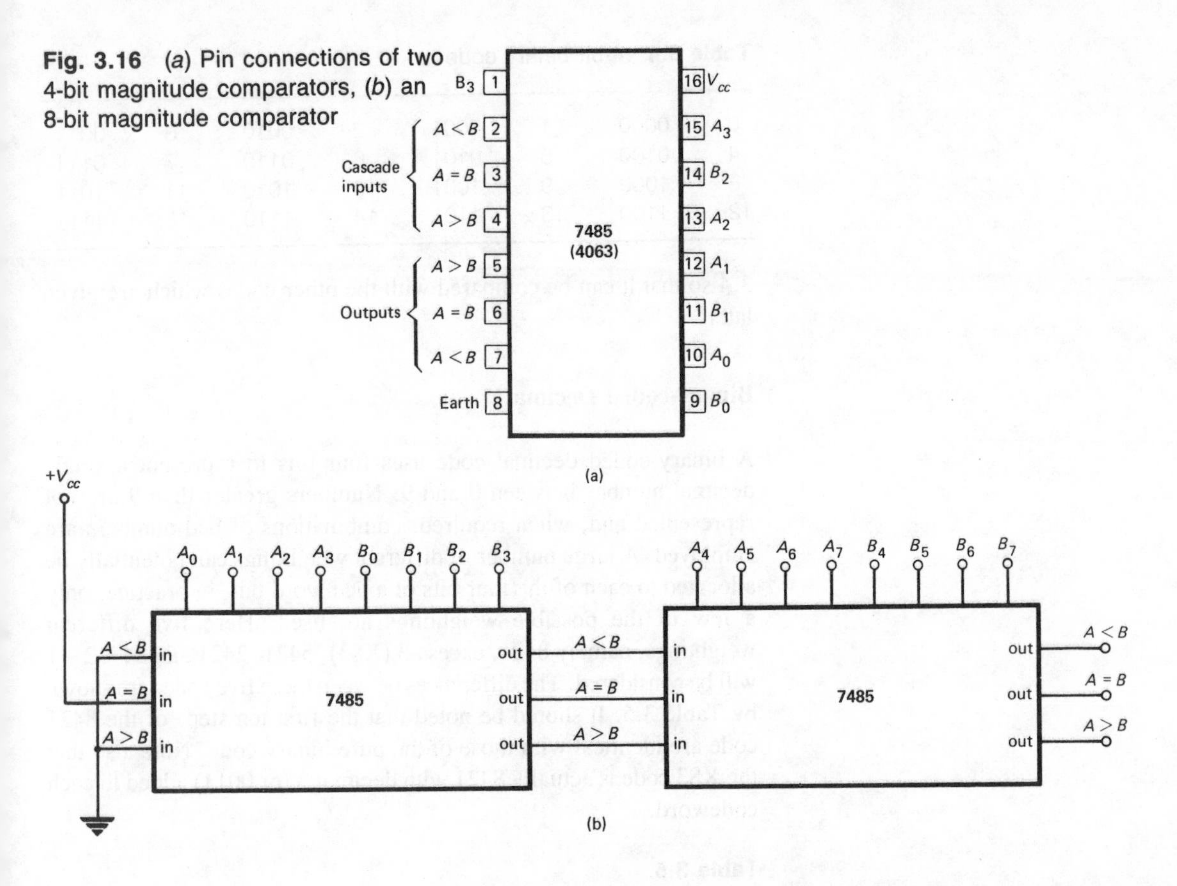

Fig. 3.16 (*a*) Pin connections of two 4-bit magnitude comparators, (*b*) an 8-bit magnitude comparator

tions and these are shown in Fig. 3.16(*a*). If the two numbers to be compared contain more than four bits, two or more ics will be needed. Figure 3.16(*b*) shows how two 7485 4-bit magnitude comparators are connected together to enable two 8-bit numbers to be compared. It can be seen that the $A>B$, $A=B$, and $A<B$ outputs of the least significant comparator (i.e. the one whose inputs are A_0, A_1, A_2, A_3 and B_0, B_1, B_2, B_3) are connected to the similarly labelled inputs of the next significant stage. The least significant stage must have its $A=B$ input terminals connected to $+V_{CC}$ volts and its $A>B$ and $A<B$ terminals connected to earth.

Codes

In digital circuitry, numbers are represented by a sequence of the binary digits or bits 1 and 0. The number of *words* or codewords which can be formed from n bits is equal to 2^n. For example, 4 bits can give 16 different words. By allocating a different value or *weighting* to each of the four bits, a large number of different codes can be generated. However, the number of codes actually employed is relatively few and only the more important of them will be discussed here.

The basic 4-bit binary code is well known but it is repeated in Table

Table 3.4 4-bit binary code

0	0000	1	0001	2	0010	3	0011
4	0100	5	0101	6	0110	7	0111
8	1000	9	1001	10	1010	11	1011
12	1100	13	1101	14	1110	15	1111

3.4 so that it can be compared with the other codes which are given later.

Binary-coded Decimal

A binary-coded decimal code uses four bits to represent a single decimal number between 0 and 9. Numbers greater than 9 are not represented and, when required, combinations of bcd numbers are employed. A large number of different weightings can potentially be allocated to each of the four bits of a bcd word but, in practice, only a few of the possible weightings are used. Here five different weightings, namely 8421, excess-3 (XS3), 5421, 2421 and 84−2−1 will be considered. The differences between these five codes are shown by Table 3.5. It should be noted that the first ten steps of the 8421 code are identical with those of the pure binary code. Note also that the XS3 code is actually 8421 with decimal 3 (or 0011) added to each codeword.

Table 3.5

	8421	*XS3*	*5421*	*2421*	*84−2−1*
0	0000	0011	0000	0000	0000
1	0001	0100	0001	0001	0111
2	0010	0101	0010	0010	0110
3	0011	0110	0011	0011	0101
4	0100	0111	0100	0100	0100
5	0101	1000	0101	0101	1011
6	0110	1001	0110	0110	1010
7	0111	1010	0111	0111	1001
8	1000	1011	1011	1110	1000
9	1001	1100	1101	1111	1111

Example 3.1

Represent the decimal number 38 in each of the bcd codes given in Table 3.5.

Solution

a) 8421 0011 1000
b) XS3 0110 1011
c) 5421 0011 1011
d) 2421 0011 1110
e) 84−2−1 0101 1000

The most frequently employed of these bcd codes is the 8421 version, partly because of its close similarity to pure binary. The 2421 code is sometimes employed in digital-to-analogue converters (Chapter 7) because it allows the use of a smaller range of resistance values.

Two features of a code that may influence its choice for a particular application are, first, any *self-complementing* property and, secondly, any *reflective*, property it may possess.

Some codes possess the advantage of being **self-complementing**. This means that their logic and arithmetic complements are identical. For example, the 9s complement of an XS3, or a 2421 bcd number, is the same as its logical complement. Some examples taken from Table 3.5 are:

a) Decimal 5 = 1000 in XS3 — 9s complement = 4 = 0111
Logical complement = 0111

b) Decimal 7 = 1010 in XS3 — 9s complement = 2 = 0101
Logical complement = 0101

This property can be of advantage in circuits which employ decimal arithmetic.

Gray Code

A code is said to be **reflective** if two numbers which are equally spaced either side of the centre numbers 7 and 8 differ in only one bit. Several examples exist but the most important, which is also an example of a *unit distance code*, is known as the **Gray code**.

Table 3.6 Gray Code

0	1	2	3	4	5	6	7	8	9
0000	0001	0011	0010	0110	0111	0101	0100	1100	1101
10	**11**	**12**	**13**	**14**	**15**				
1111	1110	1010	1011	1001	1000				

A unit distance code is one that changes in only one bit from one codeword to the next adjacent codeword. The **Gray code** is given in Table 3.6. The Gray code finds particular application in conjunction with rotational encoders which convert the angular position of a shaft into an equivalent binary number. The code may seem to be difficult to remember but a useful, easily remembered means of converting from pure binary into Gray is available.

First, put a 0 in front of the most significant bit. Then, carry out the exclusive-OR logical operation on adjacent bits starting from the left.

Example 3.2

Convert the pure binary numbers 1100 and 0111 into the Gray code.

Solution

a)
```
0 1 1 0 0
 \/\/\/\/
 1 0 1 0
```

b)
```
0 0 1 1 1
 \/\/\/\/
 0 1 0 0
```

Alphanumeric Codes

An alphanumeric code is one that includes all the letters of the alphabet, punctuation marks and arithmetic signs as well as numbers. One such code is known as the Murray code and is widely employed in telegraphy systems. The code most often used in conjunction with computers and microprocessors is known as the *American Code for Information Interchange* or **ASCII** and it is listed in Table 3.7.

Table 3.7 ASCII

	b_7			0	0	0	0	1	1	1	1
		b_6		0	0	1	1	0	0	1	1
			b_5	0	1	0	1	0	1	0	1
b_4	b_3	b_2	b_1								
0	0	0	0	NUL	DLE	SP	0	@	P	′	p
0	0	0	1	SOH	DC1	!	1	A	Q	a	q
0	0	1	0	STX	DC2	"	2	B	R	b	r
0	0	1	1	ETX	DC3	#	3	C	S	c	s
0	1	0	0	EOT	DC4	$	4	D	T	d	t
0	1	0	1	ENQ	NAK	%	5	E	U	e	u
0	1	1	0	ACK	SYN	&	6	F	V	f	v
0	1	1	1	BEL	ETB	'	7	G	W	g	w
1	0	0	0	BS	CAN	(	8	H	X	h	x
1	0	0	1	HT	EM	)	9	I	Y	i	y
1	0	1	0	LF	SUB	*	:	J	Z	j	z
1	0	1	1	VT	ESC	+	;	K	[	k	{
1	1	0	0	FF	FS	,	<	L	\	l	|
1	1	0	1	CR	GS	—	=	M	]	m	}
1	1	1	0	SO	RS	.	>	N	Λ	n	~
1	1	1	1	SI	US	/	?	O	—	o	DEL

Each character requires 7 bits. Some of the characters shown in the table are various control characters used in data systems; one example is CR which stands for carriage return.

From the table decimal 0 = 0110000 = 48_{10}
A = 1000001 = 65_{10}
j = 1101010 = 106_{10} and so on.

Code Converters

Many instances arise for a decimal number to be encoded into the corresponding binary or binary coded decimal (or some other code) number. Similarly, it is often necessary to decode a number from pure binary or bcd into decimal.

Decimal-to-Binary Coded Decimal Converter

A **decimal-to-bcd converter** will have nine input lines representing respectively the decimal integers 0, 1, 2, 3, 4, 5, 6, 7, 8, 9, and four output lines representing 2^0, 2^1, 2^2 and 2^3 respectively. The presence of any one of the nine decimal numbers is indicated by a high level (logic 1 voltage) on the appropriate input line and a low level (logic 0 voltage) on all of the remaining input lines. Decimal zero is represented by *all* the input lines being at the low level.

The converter is required to generate the logic 1 state on the appropriate output lines to produce the binary equivalent of the decimal input number. The truth table of a decimal-to-bcd converter is given by Table 3.8.

Table 3.8 Decimal-to-bcd truth table

Decimal		0	1	2	3	4	5	6	7	8	9
bcd outputs	2^0	0	1	0	1	0	1	0	1	0	1
	2^1	0	0	1	1	0	0	1	1	0	0
	2^2	0	0	0	0	1	1	1	1	0	0
	2^3	0	0	0	0	0	0	0	0	1	1

From Table 3.8 the Boolean equation describing the operation of the circuit is

$$2^0 = 1+3+5+7+9$$

$$2^1 = 2+3+6+7$$

$$2^2 = 4+5+6+7$$

$$2^3 = 8+9$$

If OR gates with two, four and five inputs are available, e.g. 4071, 4072 and 4078, the circuit given in Fig. 3.17(*a*) could be used at the expense of using three integrated circuits (the 4072 is a dual 4-input gate). The number of ics needed can be reduced to two if the 4075 triple 3-input OR gate is used instead (Fig. 3.17(*b*)).

The design of a **bcd-to-decimal converter** can also start from the

Fig. 3.17 Two decimal-to-bcd converters

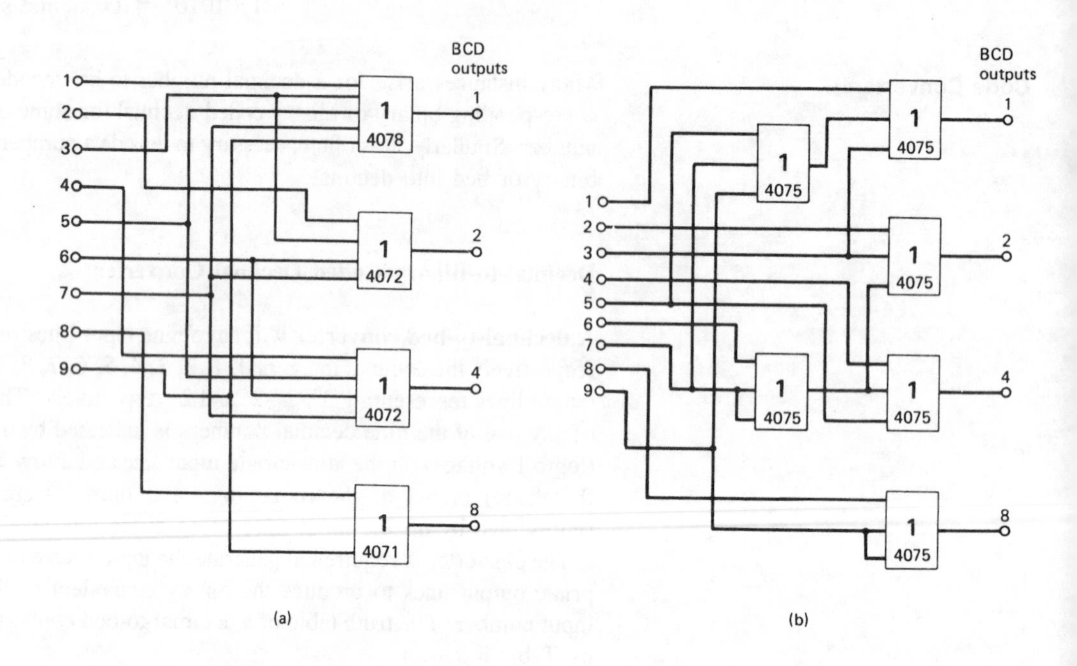

Fig. 3.18 BCD-to-decimal converter

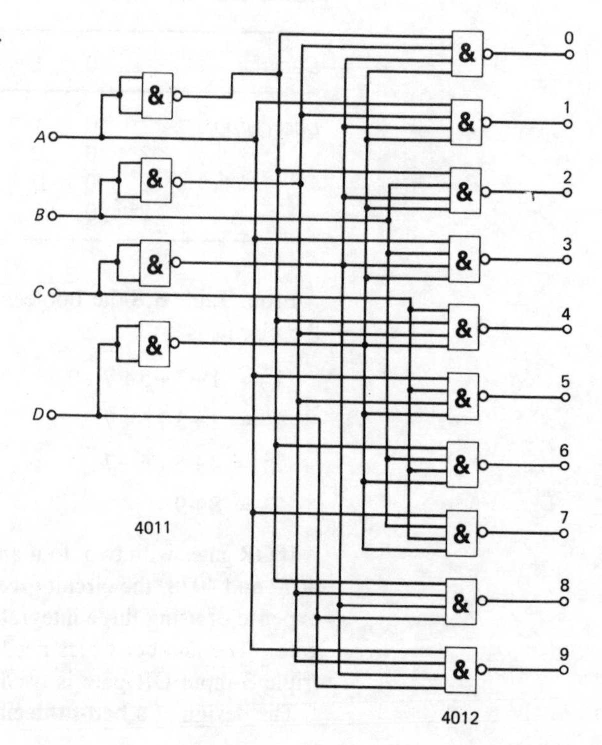

truth table given by Table 3.8. Let $2^0 = A$, $2^1 = B$, $2^2 = C$ and $2^3 = D$. Then

$$0 = \bar{A}\bar{B}\bar{C}\bar{D} \quad 1 = A\bar{B}\bar{C}\bar{D} \quad 2 = \bar{A}B\bar{C}\bar{D} \quad 3 = AB\bar{C}\bar{D} \quad 4 = \bar{A}\bar{B}C\bar{D}$$

$$5 = A\bar{B}C\bar{D} \quad 6 = \bar{A}BC\bar{D} \quad 7 = ABC\bar{D} \quad 8 = \bar{A}\bar{B}\bar{C}D \quad 9 = A\bar{B}\bar{C}D$$

These equations can be implemented directly using four inverters and ten AND gates but an alternative circuit, shown by Fig. 3.18, uses NAND gates only. In this circuit, of course, the *presence* of a given decimal number is denoted by the binary 0 state. The circuit shown could be implemented using five dual 4-input NAND gates such as the 4012 and one quad 2-input NAND gate, e.g. the 4011. Alternatively, a hex inverter (4069) could replace the 4011. If the "don't care" numbers 10 through to 15 are taken into account the circuit can be implemented using only one dual 4-input NAND gate, two triple 3-input NAND gates and one quad 2-input NAND gate.

Converters can be designed in similar manner for the conversion of a number from one code into any other code, e.g. pure binary into excess-3.

Binary-to-Gray Converters

Table 3.9

Decimal no.	Binary				Gray			
	D	C	B	A	G_4	G_3	G_2	G_1
0	0	0	0	0	0	0	0	0
1	0	0	0	1	0	0	0	1
2	0	0	1	0	0	0	1	1
3	0	0	1	1	0	0	1	0
4	0	1	0	0	0	1	1	0
5	0	1	0	1	0	1	1	1
6	0	1	1	0	0	1	0	1
7	0	1	1	1	0	1	0	0
8	1	0	0	0	1	1	0	0
9	1	0	0	1	1	1	0	1
10	1	0	1	0	1	1	1	1
11	1	0	1	1	1	1	1	0
12	1	1	0	0	1	0	1	0
13	1	1	0	1	1	0	1	1
14	1	1	1	0	1	0	0	1
15	1	1	1	1	1	0	0	0

Table 3.9 shows the binary and the Gray code equivalents of the decimal numbers 0 through to 15. From the truth table,

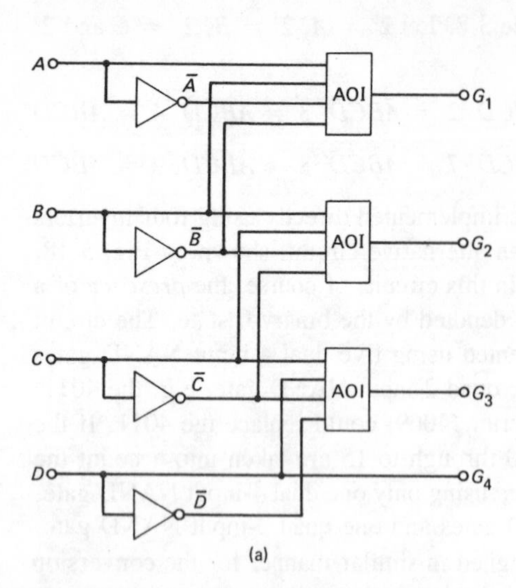

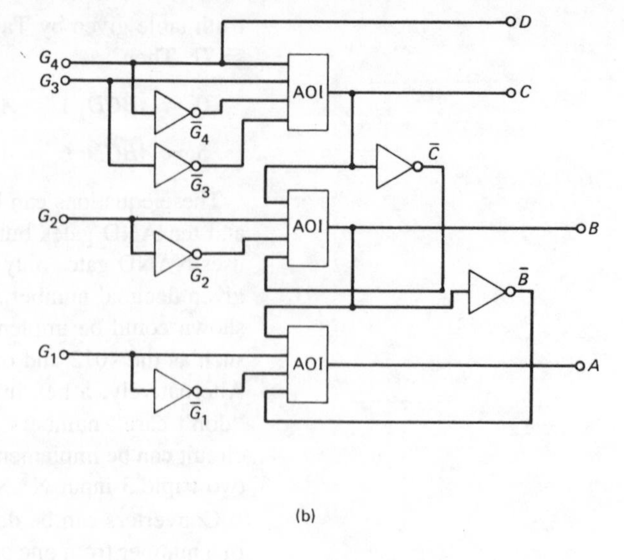

Fig. 3.19 (*a*) Binary-to-Gray code converter, (*b*) Gray-to-binary code converter

$$G_1 = A\bar{B}\bar{C}\bar{D} + \bar{A}B\bar{C}\bar{D} + A\bar{B}C\bar{D} + \bar{A}BC\bar{D} + A\bar{B}\bar{C}D + \bar{A}B\bar{C}D + A\bar{B}CD + \bar{A}BCD$$
$$G_2 = \bar{A}B\bar{C}\bar{D} + AB\bar{C}\bar{D} + \bar{A}\bar{B}C\bar{D} + A\bar{B}C\bar{D} + \bar{A}B\bar{C}D + AB\bar{C}D + \bar{A}\bar{B}CD + A\bar{B}CD$$
$$G_3 = \bar{A}\bar{B}C\bar{D} + A\bar{B}C\bar{D} + \bar{A}BC\bar{D} + ABC\bar{D} + \bar{A}\bar{B}\bar{C}D + A\bar{B}\bar{C}D + \bar{A}B\bar{C}D + AB\bar{C}D$$
$$G_4 = \bar{A}\bar{B}\bar{C}D + A\bar{B}\bar{C}D + \bar{A}B\bar{C}D + AB\bar{C}D + \bar{A}\bar{B}CD + A\bar{B}CD + \bar{A}BCD + ABCD$$

Mapping on a Karnaugh map and simplifying gives

$$G_1 = \bar{A}B + A\bar{B} \quad G_2 = \bar{B}C + B\bar{C} \quad G_3 = \bar{C}D + C\bar{D} \quad G_4 = D$$

There are several ways in which the terms for G_1, G_2, G_3 and G_4 can be implemented.

1 Using three 2-wide 2-input AOI gates and four inverters as shown by Fig. 3.19(*a*). This implementation could be achieved with three ics; for example two dual 2-wide 2-input AOI gates such as the 7450 and one hex inverter such as the 7404. It is left as an exercise to confirm the operation of this circuit.

2 Using three exclusive-OR circuits and a hex inverter.

3 Using NAND or NOR gates or, of course, a combination of AND and OR gates and inverters.

A **Gray-to-binary converter** can be similarly designed by reading the truth table of Table 3.9 the other way around. Thus

$$A = G_1\bar{G}_2\bar{G}_3\bar{G}_4 + \bar{G}_1G_2\bar{G}_3\bar{G}_4 + G_1G_2G_3\bar{G}_4 + \bar{G}_1\bar{G}_2G_3\bar{G}_4$$
$$+ G_1\bar{G}_2G_3G_4 + \bar{G}_1G_2G_3G_4 + G_1G_2\bar{G}_3G_4 + \bar{G}_1\bar{G}_2\bar{G}_3G_4 \quad (3.6)$$

This equation cannot be simplified by mapping.

$$B = G_1 G_2 \bar{G}_3 \bar{G}_4 + \bar{G}_1 G_2 \bar{G}_3 \bar{G}_4 + G_1 \bar{G}_2 G_3 \bar{G}_4 + \bar{G}_1 \bar{G}_2 G_3 \bar{G}_4$$
$$+ G_1 G_2 G_3 G_4 + \bar{G}_1 G_2 G_3 G_4 + G_1 \bar{G}_2 \bar{G}_3 G_4 + \bar{G}_1 \bar{G}_2 \bar{G}_3 G_4 \quad (3.7)$$

$$C = \bar{G}_1 G_2 G_3 \bar{G}_4 + G_1 G_2 G_3 \bar{G}_4 + G_1 \bar{G}_2 G_3 \bar{G}_4 + \bar{G}_1 \bar{G}_2 G_3 \bar{G}_4$$
$$+ \bar{G}_1 G_2 \bar{G}_3 G_4 + G_1 G_2 \bar{G}_3 G_4 + G_1 \bar{G}_2 \bar{G}_3 G_4 + \bar{G}_1 \bar{G}_2 \bar{G}_3 G_4 \quad (3.8)$$

Mapping and simplifying,

$$C = G_3 \bar{G}_4 + \bar{G}_3 G_4 \quad (3.9)$$

$$B = G_2 G_3 G_4 + \bar{G}_2 G_3 \bar{G}_4 + G_2 \bar{G}_3 \bar{G}_4 + \bar{G}_2 \bar{G}_3 G_4 \quad (3.10)$$

$$D = G_4$$

The equations for A, B, and C can be implemented directly but considerable simplification results if it is noticed that

$$B = \bar{G}_2 C + G_2 \bar{C} \quad (3.11)$$

and

$$A = G_1 \bar{B} + \bar{G}_1 B \quad (3.12)$$

The AOI implementation of the Gray-to-binary converter is shown in Fig. 3.19(*b*).

The ttl logic family includes some converters:

7442 bcd-to-decimal converter
7443 excess 3-to-decimal converter
74145 bcd-to-decimal converter
74184 bcd-to-pure binary converter
74185 binary-to-bcd converter.

Multiplexers and Demultiplexers

Two other combinational logic circuits often used in digital circuitry are known as the *multiplexer* or *data selector* and the *demultiplexer* or *decoder*. The principle of operation of a multiplexer is illustrated by Figs. 3.20(*a*) and (*b*).

The **multiplexer** has a number n of inputs, any one of which can be switched to the output terminal under the instruction of the select S control input. The idea is shown more clearly by Fig. 3.20(*b*). Only one of the switches can be closed at a time and this switches the appropriate input to the output.

Consider the design of a simple 2-to-1 line multiplexer. When the select signal S is at the logical 0 level, the input D_0 should be connected to the output. Conversely, when $S = 1$, the input D_1 should be connected to the output. The truth table describing this operation is given by Table 3.10 and the required circuit is shown by Fig. 3.21.

As another simple example, consider a 4-to-1 multiplexer. The multiplexer circuit must have two inputs A and B since $2^2 = 4$ and a suitable circuit is shown in Fig. 3.22. Each AND gate has three inputs, one of which is connected to a data input line, and the other two are connected to the selection circuit which merely consists of four inverters. Suppose the input address is $A = 0$ and $B = 1$. Then $\bar{A} = 1$ and $\bar{B} = 0$ and only the third AND gate from the top is enabled.

Fig. 3.20 The principle of a multiplexer

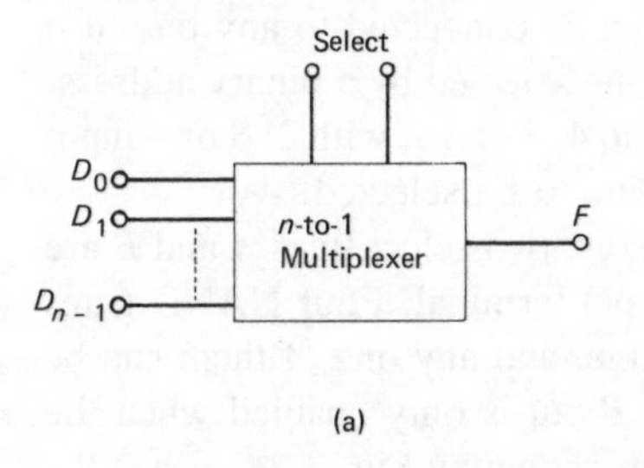

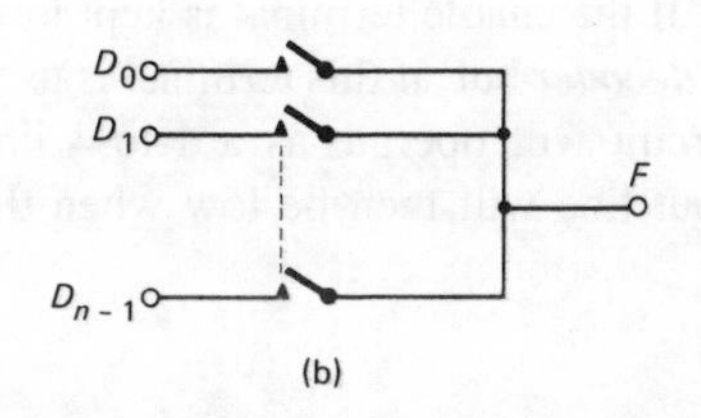

Table 3.10 2-to-1 line multiplexer

S	D_0	D_1	F
0	0	0	0
0	0	1	0
0	1	0	1
0	1	1	1
1	0	0	0
1	0	1	1
1	1	0	0
1	1	1	1

Fig. 3.21 2-to-1 line multiplexer

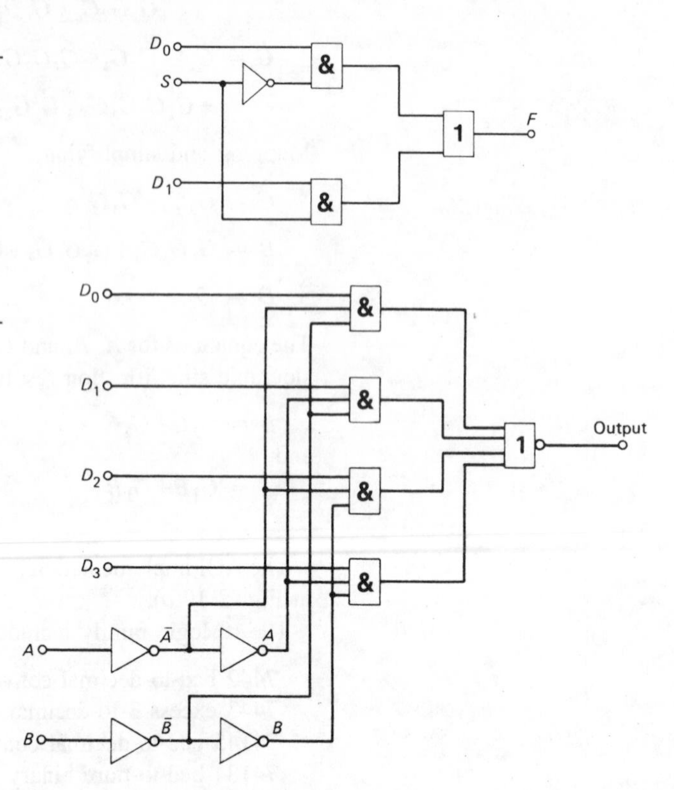

Fig. 3.22 4-to-1 line multiplexer

The output of the circuit is then D_2.

TTL multiplexers include the 74151 8-to-1 line, the 74157/8 quad 2-to-1 line, the 74153 dual 4-to-1 line, and the 74150 16-to-1 line.

A **demultiplexer** or **decoder** performs the inverse function to that performed by a multiplexer. The basic concept of a decoder is shown in Fig. 3.23. A single input line can be connected to any one of n output lines, the required output being selected by a binary address. The number of output lines is equal to 4, 8 or 16, with 2, 3 or 4 input address bits needed to allow any line to be selected.

Consider a 1-to-4 line demultiplexer; two select lines A and B are needed and an *enable* or *data-in* input terminal. Four NAND gates are required, one for each output line, and any one of them can be selected by the address bits A and B but is only enabled when the enable terminal is low. The circuit is shown by Fig. 3.24. When the enable terminal is low, the selected NAND gate will have a low output; all the other outputs will be high. If the enable terminal is kept low, the circuit will act as a 1-to-4 line *decoder* but, if this terminal is used as a data input terminal, the circuit will operate as a 1-to-4 line demultiplexer. The selected output line will then be low when the

Fig. 3.23 The principle of a demultiplexer or decoder

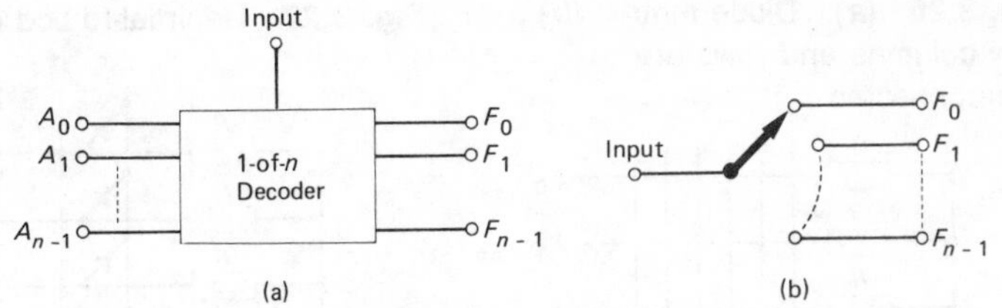

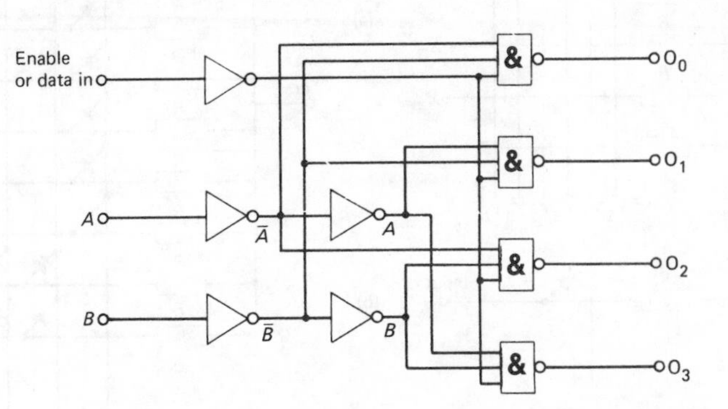

Fig. 3.24 Decoder/demultiplexer

data input is low and high when the data input is high. The other outputs will remain high.

A number of decoders/demultiplexers are included in the ttl family: the 74138 3-to-8 line, the 74139 dual 2-to-4 line, and the 74154 4-to-16 line.

Diode Matrices

The implementation of a code converter or a decoder can also be achieved using a **diode matrix** or read only memory (ROM). The Boolean equation to be implemented must first be expressed in its canonical form. The block diagram of a ROM code converter is shown in Fig. 2.35. The *m*-bit input word is decoded to give 2^m word lines and then each word line is encoded into the wanted *n*-bit output word. The numbers *m* and *n* are not necessarily different from one another.

Fig. 3.25 ROM code converter

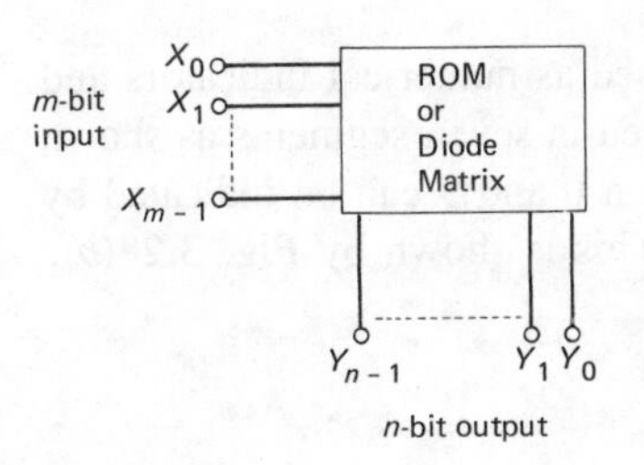

The diode matrix consists of a number *m* of rows and another number *n* of columns arranged in the manner indicated by Fig. 3.26(*a*).

Suppose a diode is connected between the intersection of a row and a column as shown by Fig. 3.26(*b*). The column line will take up either the logical 1 or the logical 0 level depending on the logical state of the row line. If the row input is held at logical 0, current will flow and V_{CC} volts will be dropped across the resistor *R*. The diode will then be non-conducting and the column line will be low. Conversely, if the row input is high, no voltage will be dropped across resistor *R* and the diode will turn ON. The column line is then high.

The design of a diode matrix code converter (or other circuit) therefore consists of the connection of diodes between rows and columns at appropriate points in the matrix to obtain the required output logic.

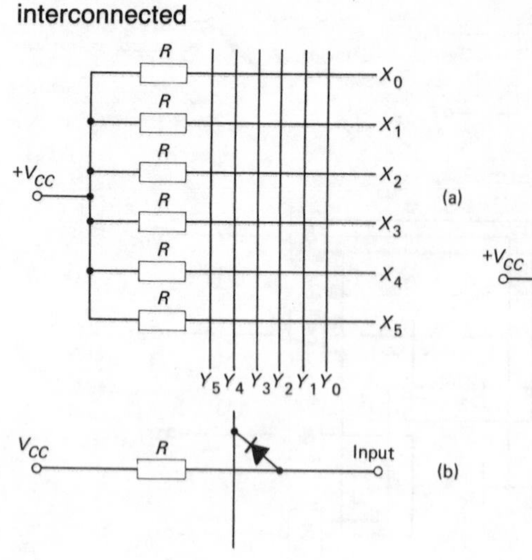

Fig. 3.26 (*a*) Diode matrix, (*b*) how columns and rows are interconnected

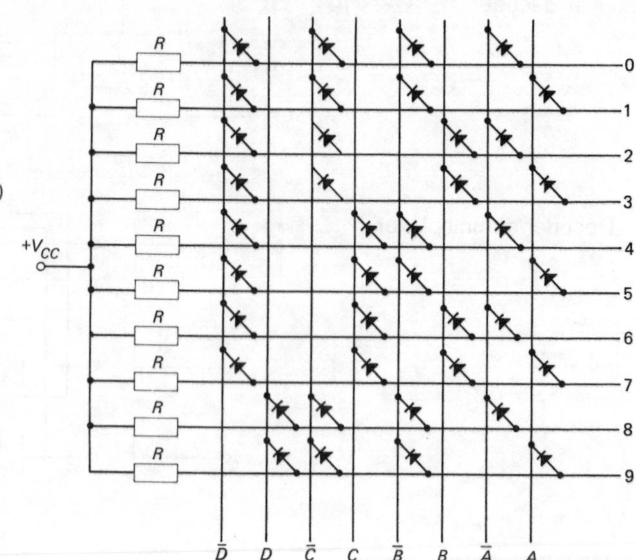

Fig. 3.27 Decimal-to-bcd converter

Suppose a decimal-to-bcd converter is to be designed. Ten rows and eight columns are necessary (see Fig. 3.27). When all ten of the decimal inputs are low, none of the diodes will conduct and all the outputs are low. If, say, the decimal 5 line goes high, no current then flows in that particular row and the columns A, $\bar{B}$, C and $\bar{D}$ go high. This is, of course, the 8421 equivalent of decimal 5.

Seven-segment Displays

7-segment displays are generally used as numerical indicators and consist of a number of LEDs arranged in seven segments as shown in Fig. 3.28(*a*). Any number between 0 and 9 can be indicated by lighting the appropriate segments. This is shown by Fig. 3.28(*b*).

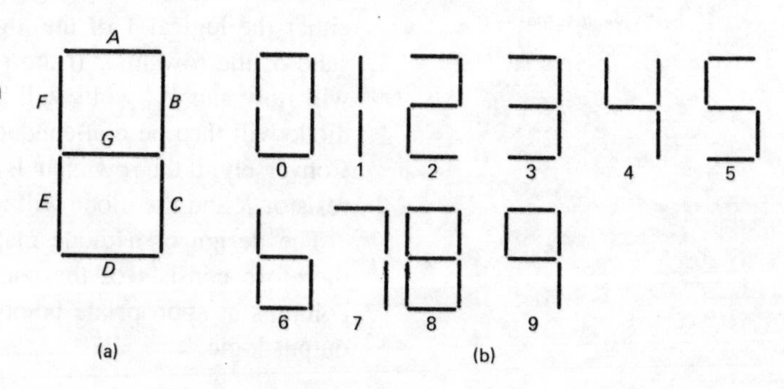

Fig. 3.28 7-segment display: (*a*) arrangement of LEDs, (*b*) indication of numbers 0 to 9

Table 3.11 7-segment display

Decimal number displayed	*Inputs*				*Outputs*						
	D	*C*	*B*	*A*	*a*	*b*	*c*	*d*	*e*	*f*	*g*
0	0	0	0	0	1	1	1	1	1	1	0
1	0	0	0	1	0	1	1	0	0	0	0
2	0	0	1	0	1	1	0	1	1	0	1
3	0	0	1	1	1	1	1	1	0	0	1
4	0	1	0	0	0	1	1	0	0	1	1
5	0	1	0	1	1	0	1	1	0	1	1
6	0	1	1	0	0	0	1	1	1	1	1
7	0	1	1	1	1	1	1	0	0	0	0
8	1	0	0	0	1	1	1	1	1	1	1
9	1	0	0	1	1	1	1	0	0	1	1

The truth table of a 7-segment display is given by Table 3.11; note that numbers greater than 0 do not appear in the display and so they are "don't cares".

From the truth table a number of segment maps can be obtained, each of which maps the inputs which must be high for a segment to be illuminated.

Thus segment a is ON when the input decimal number is 0, 2, 3, 5, 7, 8 or 9. Hence

$$a = \bar{A}\bar{B}\bar{C}\bar{D}+\bar{A}B\bar{C}\bar{D}+AB\bar{C}\bar{D}+A\bar{B}C\bar{D}+ABC\bar{D}+\bar{A}\bar{B}\bar{C}D+A\bar{B}\bar{C}D$$

Mapping

	A		$\bar{A}$		
C	X	X	X	X	D
C	1	1	0	0	$\bar{D}$
$\bar{C}$	1	0	1	1	$\bar{D}$
$\bar{C}$	X	1	1	X	D
	B	$\bar{B}$	$\bar{B}$	B	

From the map

$$a = \bar{A}\bar{C}+AB+AC+D$$

Similarly for the other segments:

	A	A	$\bar{A}$	$\bar{A}$	
C	X	X	X	X	D
C	1	0	1	0	$\bar{D}$
$\bar{C}$	1	1	1	1	$\bar{D}$
$\bar{C}$	X	1	1	X	D
	B	$\bar{B}$	$\bar{B}$	B	

	A	A	$\bar{A}$	$\bar{A}$	
C	X	X	X	X	D
C	1	1	1	1	$\bar{D}$
$\bar{C}$	1	1	1	0	$\bar{D}$
$\bar{C}$	X	1	1	X	D
	B	$\bar{B}$	$\bar{B}$	B	

	A	A	$\bar{A}$	$\bar{A}$	
C	X	X	X	X	D
C	0	1	0	1	$\bar{D}$
$\bar{C}$	1	0	1	1	$\bar{D}$
$\bar{C}$	X	0	1	X	D
	B	$\bar{B}$	$\bar{B}$	B	

	A	A	$\bar{A}$	$\bar{A}$	
C	X	X	X	X	D
C	0	0	0	1	$\bar{D}$
$\bar{C}$	0	0	1	1	$\bar{D}$
$\bar{C}$	X	0	1	X	D
	B	$\bar{B}$	$\bar{B}$	B	

	A	A	$\bar{A}$	$\bar{A}$	
C	X	X	X	X	D
C	0	1	1	1	$\bar{D}$
$\bar{C}$	0	0	1	0	$\bar{D}$
$\bar{C}$	X	1	1	X	D
	B	$\bar{B}$	$\bar{B}$	B	

	A	A	$\bar{A}$	$\bar{A}$	
C	X	X	X	X	D
C	0	1	1	1	$\bar{D}$
$\bar{C}$	1	0	0	1	$\bar{D}$
$\bar{C}$	X	1	1	X	D
	B	$\bar{B}$	$\bar{B}$	B	

From the maps,

$$b = AB + \bar{A}\bar{B} + \bar{C}$$
$$c = A + \bar{B} + C$$
$$d = \bar{A}B + \bar{A}\bar{C} + B\bar{C} + A\bar{B}C$$
$$e = \bar{A}B + \bar{A}\bar{C}$$
$$f = \bar{A}\bar{B} + \bar{B}C + \bar{A}C + D$$
$$g = \bar{A}B + \bar{B}C + B\bar{C} + D$$

This logic can be implemented in one of the various ways described earlier.

A typical 7-segment display is manufactured in a 14-pin dil package with the cathode of each LED being brought out to a terminal together with the common anode.

Clearly, the 7-segment display needs a 7-bit input signal and so a *decoder* is required to convert the digital signal to be displayed into the corresponding 7-segment signal. Decoder/driver circuits *can* be made using ssi devices but more usually a purpose-built ic would be used. A commonly employed device is the ttl 7447 bcd-to-7-segment decoder/driver. This ic has four input pins A, B, C and D to which the bcd input signal is applied and seven output pins, labelled as *a* through to *g*. When an output is high the segment to which it is connected lights. If, for example, the input signal is 0101, outputs a, f, g, c, and d go high so that the decimal number 5 is illuminated.

The ic includes a facility known as *remote blanking*. Two other pins, labelled as RB_{in} and RB_{out} are provided. If the RB_{in} pin of the most significant 7-segment display is earthed *and* inputs A, B, C and D are all low, then its RB_{out} pin will be low and so are all segment outputs. The RB_{out} pin of each display is connected to the RB_{in} pin of the next most significant 7-segment display. This connection ensures that leading 0s in a displayed decimal number are not visible, e.g. for a 4-bit display 617 would be displayed and not 0617.

Other bcd-to-7-segment decoders are available in the ttl logic family, such as the 7446/8/9 and the 74246/7/8/9, while the cmos family includes the 4511.

When a number of digits are to be displayed, as in a digital meter for example, time division multiplexing is often used to reduce the power consumption. With tdm each digit of the display is only energized for a fraction of the time the number is displayed. In a 4-digit display, for example, each digit is energized for 1/4th of the time. The disadvantage of the tdm system is the extra circuitry that is needed.

Exercises 3

3.1 Design a logic circuit to compare two 3-bit binary numbers A and B. The circuit should give outputs to indicate each of the conditions $A>B$, $A = B$ and $A<B$.

3.2 Design, with the aid of a truth table and a Karnaugh map, a majority decision circuit that will give an output at logical 1 whenever two out of its three inputs are at 1.

3.3 State what is meant by a diode matrix. Draw a diode matrix to provide the outputs listed below when inputs A, B and C and their complements are applied.

$$ABC,\ AB\bar{C},\ A\bar{B}C,\ \bar{A}BC,\ A\bar{B}\bar{C},\ \bar{A}B\bar{C},\ \bar{A}\bar{B}C$$

3.4 Design a logic circuit to have the following characteristics: inputs A, B and C representing, in binary, the denary numbers 0 through to 7, where A is the least significant bit. There should be two outputs X and Y; X should be at 1 only when the input is an even number but not zero; Y should be at 1 only when the input is an odd number. Draw your circuit using either NAND or NOR gates only.

Table 3.12

Input denary numbers	0	1	2	3	4	5	6	7
Output denary numbers	0	6	4	7	0	3	7	7

3.5 Design a logic circuit that will change the input denary numbers given in Table 3.12 into the given output denary numbers.
Both input and output numbers are represented in binary code.

3.6 Draw a diode matrix suitable for converting 4-bit binary into hexadecimal.

3.7 Show that the circuit given in Fig. 3.19 generates the function

$$G_1 = \bar{A}B + A\bar{B},\ G_2 = \bar{B}C + B\bar{C},\ G_3 = \bar{C}D + C\bar{D},\ \text{and}\ G_4 = D.$$

3.8 Design a circuit to give the product of two binary numbers A and B where A is in the range 0–3 and B is in the range 0–5.

3.9 Show that the design of a bcd-to-XS3 converter is satisfied by the equations

$$B_0 = \bar{A}_0 \qquad B_1 = A_0A_1 + \bar{A}_0\bar{A}_1 \qquad B_2 = \bar{A}_0\bar{A}_1A_2 + A_1\bar{A}_2 + A_0\bar{A}_2$$

$$B_3 = A_3 + A_1A_2 + A_0A_2$$

3.10 Distinguish between weighted and unit-distance codes, giving an example of each. Show how a unit-distance code can be generated using a Karnaugh map.
Devise a circuit to convert from the unit-distance code shown in Table 3.13 to the first ten states of a binary count. Simplify the circuits as far as possible, assuming that the states shown are the only permissible ones.

Table 3.13

	0	1	2	3	4	5	6	7	8	9
A	0	0	0	0	0	1	1	1	1	1
B	0	1	1	1	0	0	1	1	1	0
C	0	0	1	1	1	1	1	1	0	0
D	1	1	1	0	0	0	0	1	1	1

3.11 Design a circuit which will convert the pure binary number *ABCD* into the excess-three coded number *WXYZ*.

3.12 Draw the block diagram of a binary multiplier which uses a shift-and-add technique to multiply two *n*-bit numbers together. Explain the operation of the multiplier.

4 Flip-Flops

Introduction

The *bistable multivibrator*, or **flip-flop**, is a circuit that has two stable states. It is able to remain in either state for an indefinite period of time. The flip-flop will change state only when a trigger pulse is applied to one of its input terminals. Once the circuit has changed state, it will remain in its new state until another trigger pulse is applied to the circuit to make it revert back to its original state. A bistable multivibrator has two output terminals that are usually labelled as Q and $\bar{Q}$ since the logical state of one output terminal is *always* the complement of the logical state of the other terminal. When the circuit is in the state $Q = 1$, $\bar{Q} = 0$ it is said to be SET; conversely, when $Q = 0$ and $\bar{Q} = 1$ the circuit is said to be RESET.

Four types of flip-flop are available, known respectively as the S-R, the J-K, the D, and the T flip-flops. The S-R and J-K flip-flops can act as 1-bit stores and may be interconnected to provide various other more complex functions, such as counters and registers. The D flip-flop is used mainly to provide a time delay equal to the periodic time of the *clock* waveform. It can, however, also be connected to form counter circuits, etc. Finally, the T flip-flop acts as a *toggle*, that is it changes state every time the clock waveform is at logical 1.

The various kinds of flip-flop *can* be constructed by the suitable interconnectcion of two or more NOR or NAND gates, but it is more often convenient to use one of the integrated circuit versions that are readily available in the various logic families.

The S-R Flip-Flop

The **S-R flip-flop** has two input terminals labelled as S (for Set) and R (for Reset) and two output terminals labelled as Q and $\bar{Q}$. An S-R flip-flop may be operated either asynchronously or synchronously (i.e. clocked). When the circuit is not clocked it is often known as a *latch*.

The symbol for an S-R flip-flop is given in Fig. 4.1 and its truth table by Table 4.1. In the table the symbol Q represents the *present* state of the Q output terminal, and Q^+ represents the *next* state of the terminal *after* a set (S) or reset (R) pulse has been applied to the appropriate input terminal.

If the inputs S and R are both at logical 0, the output state of the circuit will not change and the flip-flop can be said to have stored one bit of information.

When $S = 1$ and $R = 0$, the flip-flop will change state if $Q = 0$ to $Q^+ = 1$ but it will not change state if $Q = 1$. Conversely, if $Q = 1$, a reset pulse, i.e. $R = 1$, $S = 0$, will cause the circuit to

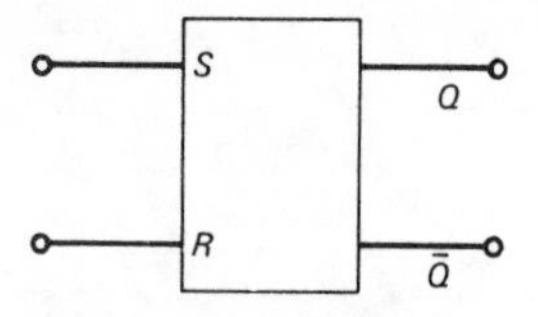

Fig. 4.1 Symbol for an S-R flip-flop

Table 4.1 S-R flip-flop truth table

S	R	Q	Q^+	
0	0	0	0	no change in state
0	0	1	1	
1	0	0	1	set operation
1	0	1	1	
0	1	0	0	reset operation
0	1	1	0	
1	1	0	X	indeterminate operation
1	1	1	X	

switch to $Q^+ = 0$, but the $R = 1$, $S = 0$ input condition will have no effect on the circuit if $Q = 0$ and $\bar{Q} = 1$.

Thus, $S = 1$, $R = 0$ will always produce the state $Q^+ = 1$; and $S = 0$, $R = 1$ will always give $Q^+ = 0$ regardless of the original state of Q.

Lastly, if $R = S = 1$ the flip-flop may or may not change state and the operation of the circuit is said to be *indeterminate*. This condition is denoted in the table by X.

The entries in the truth table can be used to derive the *state table* of the S-R flip-flop. The state table provides an alternative method of representing the operations of the flip-flop. The rows in the state table represent the output states of the circuit, while the columns correspond to the possible input states. An entry in a particular position in the state table represents the *next* state Q^+ of the output of the flip-flop for the circuit change, or *transition*, caused by applying the input shown by that column when the circuit is in the state represented by that row.

The state table of an S-R flip-flop is

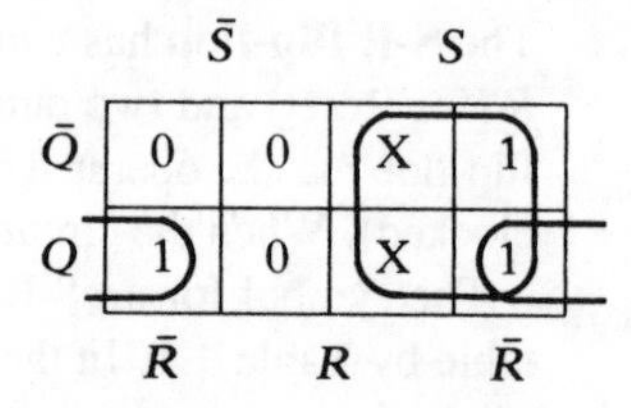

From the state table the *characteristic equations* of the S-R flip-flop can be written down:

$$Q^+ = S + Q\bar{R} \tag{4.1}$$

$$\overline{Q^+} = R + \bar{Q}\bar{S} \tag{4.2}$$

The characteristic equations can be used to derive the NOR and the NAND gate implementations of the S-R flip-flop.

Table 4.2 S-R flip-flop transition table

Present state	*Desired next state*	*Required inputs*	
Q	Q^+	S	R
0	0	0	X
0	1	1	0
1	0	0	1
1	1	X	0

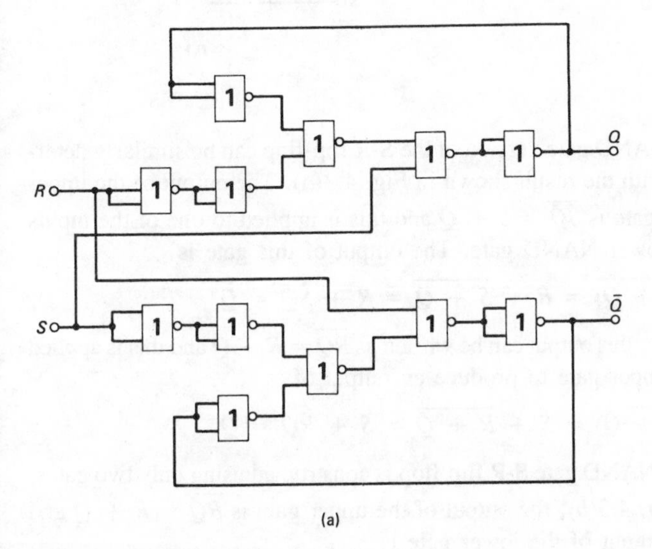

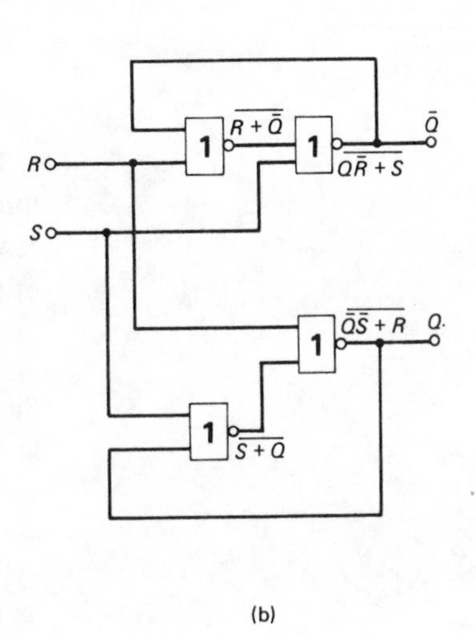

Fig. 4.2 Steps in the derivation of a circuit for the NOR gate version of an S-R flip-flop

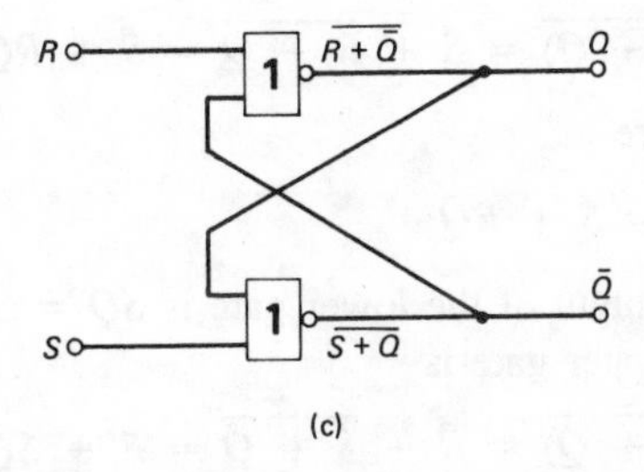

The *transition table* of a flip-flop specifies the required inputs for every possible combination of present and next states of the flip-flop. The transition table of an S-R flip-flop is given by Table 4.2.

The NOR gate implementation of the characteristic equations is shown in Fig. 4.2(*a*). This circuit can be simplified by the removal of the redundant gates to produce the arrangement shown in Fig. 4.2(*b*). The two outputs of the circuit are $\overline{Q\bar{R} + S}$ and $\overline{\bar{Q}\bar{S} + R}$ respectively or, from equations (4.1) and (4.2), they are $\bar{Q}^+$ and Q^+. If now it is realized that the outputs of the left-hand gates, i.e. $\overline{R+\bar{Q}}$ and $\overline{S+Q}$, are equal to Q^+ and $\bar{Q}^+$ respectively, i.e.

$$\overline{R + \bar{Q}} = \overline{R + R^+\bar{Q}\bar{S}} = \overline{R + \bar{Q}\bar{S}} = \overline{\bar{Q}^+} = Q^+$$

the circuit can be still further reduced to give Fig. 4.2(*c*).

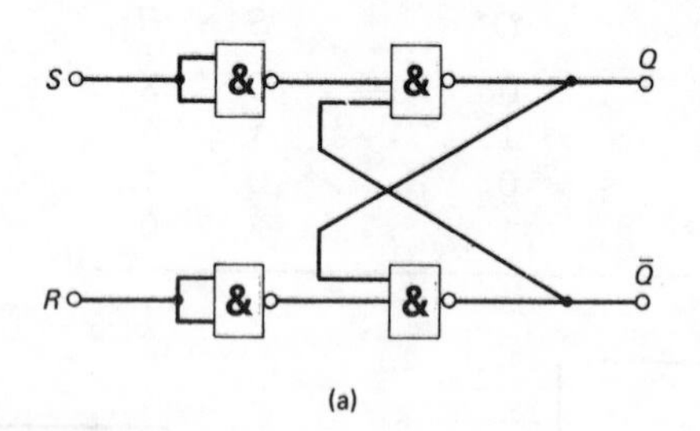

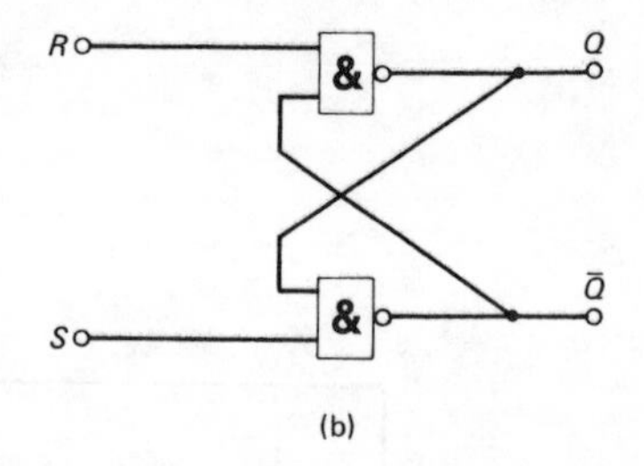

Fig. 4.3 (*a*) NAND gate S-R flip-flop, (*b*) alternative S-R flip-flop in which the indeterminate state is $S = R = 0$

The NAND gate version of the S-R flip-flop can be similarly determined with the result shown in Fig. 4.3(*a*). The output of the upper NAND gate is $\overline{\bar{S}\bar{Q}} = S + Q$ and this is applied to one of the inputs of the lower NAND gate. The output of this gate is

$$\overline{\bar{R}(S + Q)} = R + \overline{S + Q} = R + \bar{S}\bar{Q} = \overline{Q^+}$$

Similarly, this output can be written as $\overline{\bar{R}Q} = R + \bar{Q}$ and this is applied to the upper gate to produce an output of

$$\overline{\bar{S}(R + \bar{Q})} = S + \overline{R + \bar{Q}} = S + \bar{R}Q = Q^+$$

If the NAND gate S-R flip-flop is constructed using only two gates, as in Fig. 4.3(*b*), the output of the upper gate is $\overline{R\bar{Q}} = \bar{R} + Q$ and so the output of the lower gate is

$$\overline{S(\bar{R} + Q)} = \bar{S} + \overline{\bar{R} + Q} = \bar{S} + R\bar{Q}$$

Therefore

$$\overline{Q^+} = \bar{S} + R\bar{Q} \qquad (4.3)$$

The output of the lower gate is $\overline{SQ} = \bar{S} + \bar{Q}$ and so the output of the upper gate is

$$\overline{R(\bar{S} + \bar{Q})} = \bar{R} + \overline{\bar{S} + \bar{Q}} = \bar{R} + SQ$$

Therefore

$$Q^+ = \bar{R} + SQ \qquad (4.4)$$

Comparing equations (4.3) and (4.4) with equations (4.1) and (4.2) it can be seen that the S and R signals are inverted. Equations (4.3) and (4.4) relate to the truth table given by Table 4.3. This table represents a circuit whose indeterminate state is $S = R = 0$ and for which the input conditions $S = R = 1$ produce no change in the state of the circuit.

Table 4.3

S	0	0	1	1	0	0	1	1
R	0	0	0	0	1	1	1	1
Q	0	1	0	1	0	1	0	1
Q^+	X	X	1	1	0	0	0	1

The Clocked S-R Flip-Flop

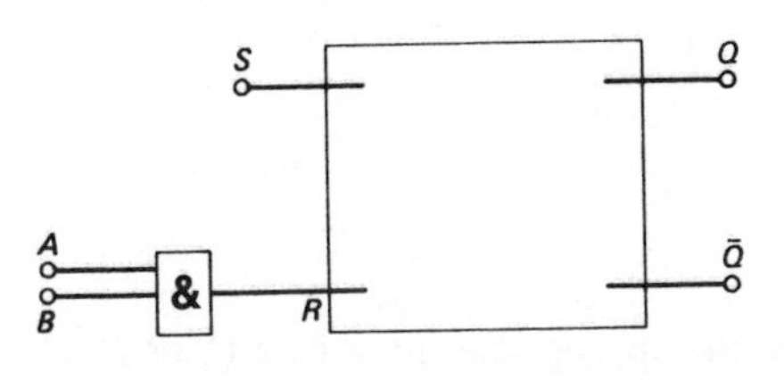

Fig. 4.4 S-R flip-flop with R input gated

A flip-flop is not generally used on its own but as part of a much larger digital system. Suppose, as an example, that the input to the reset terminal is provided by the output of a 2-input AND gate as shown in Fig. 4.4.

Consider a time when the circuit is set, $Q = 1$, $\bar{Q} = 0$, and $A = 1$, $B = 0$, so that $R = 0$. The circuit will remain set whether S is 0 or 1. If $S = 0$, the circuit will reset when $A = B = 1$. A problem can arise if the inputs A and B are supposed to change their states simultaneously to the state $A = 0$, $B = 1$; this would ensure that R remains at 0 so that the flip-flop remains set. It is difficult, however, to ensure that A and B change state simultaneously and the possibility exists that B may change state to 1 *before* A changes state from 1 to 0. Should this situation occur, R will be at 1 for a short period of time and the flip-flop will reset, $Q = 0$, $\bar{Q} = 1$, and will remain reset when A changes to 0.

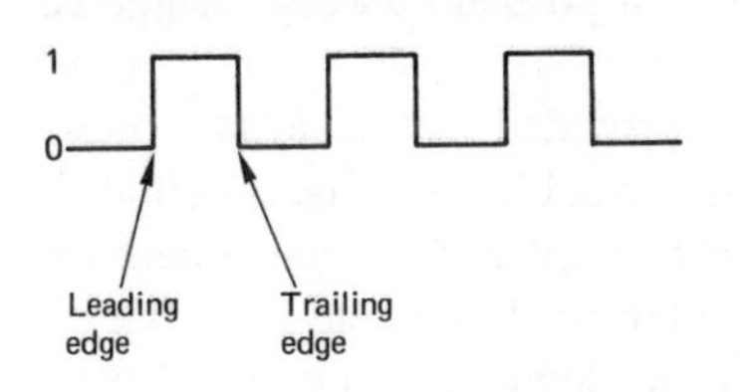

Fig. 4.5 Clock waveform

To overcome this kind of difficulty a **clocked S-R flip-flop** can be used. With such a circuit the S and R input states determine the final state of the flip-flop *but* the time at which any change in state takes place is determined by the clock.

The clock is a square waveform produced by a stable source, such as a crystal oscillator, and ideally is as shown in Fig. 4.5. The function of the clock is to ensure that all the circuits controlled, or *synchronized*, by the clock operate at the same instant in time. The clock frequency will determine the speed of operation of the system. Some circuits are designed to operate when the clock waveform is at the logical 1 voltage level; others operate at either the leading edge, or at the trailing edge, of the clock waveform.

Circuits and systems that operate under the control of the clock are said to be *synchronous*. Conversely, a non-synchronous circuit operates immediately the appropriate signal is applied to the circuit.

Fig. 4.6(a) shows how a NAND gate S-R flip-flop can be clocked. Whenever the clock is at logical 0, the output of both the input NAND gates must be at logical 1 whatever the logical states of the S and R inputs. Suppose the circuit is set; then $Q = 1$, $\bar{Q} = 0$ and so the upper right-hand gate has one input at 1 ($\bar{C}S + C\bar{S}$) and one at 0 ($\bar{Q}$), and hence the Q output remains at 1. The lower right-hand gate has both of its inputs at 1 and so the $\bar{Q}$ output remains at 0. Only

Fig. 4.6 Clocked S-R flip-flop

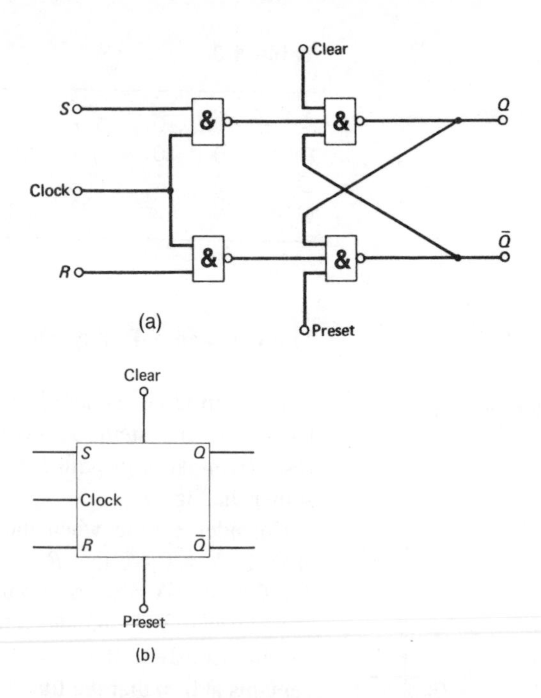

when the clock input is 1 can the appropriate input gate (S or R = 1) have an input at 0 for a possible switching action to be initiated. This means that, when the clock is 1, the circuit operates in the same way as the NAND circuit of Fig. 4.3(*a*).

The circuit is vulnerable to changes in the S and/or R inputs while the clock is high. Should this present a problem an edge-triggered device must then be used.

The clocked S-R flip-flop is often provided with clear and preset terminals (Fig. 4.6(*a*)) that allow the normal inputs to be overridden. These inputs are always non-synchronous even if the main operation is clocked. Normally the clear/preset terminals are held at the 0 logic level. The symbol for a clocked S-R flip-flop is shown in Fig. 4.6(*b*).

Switch Debouncing

When an electrical switch is operated it will generally bounce off its new contact several times (without retouching the original contact) before it settles down at its new position. In many cases such *contact bounce* is unimportant but when its presence is undesirable it can be eliminated by the use of an S-R flip-flop.

A circuit needing debouncing is shown in Fig. 4.7. When the switch operates, contact bounce produces a few blips of output voltage before a steady voltage is obtained. With the S-R flip-flop debouncing circuit of Fig. 4.7(*b*) the operation of the switch sets the flip-flop immediately. Its S terminal is taken high and the output voltage goes to, and remains at, +5 V.

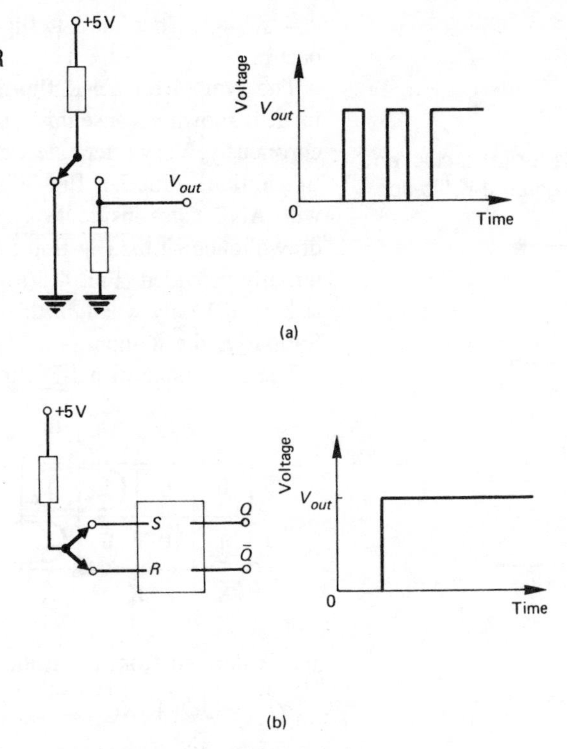

Fig. 4.7 Switch contact debouncing: (*a*) basic circuit, (*b*) S-R flip-flop connected as a debouncer

(*Note*: a Schmitt trigger can also be used as a contact debouncer.)

The J-K Flip-Flop

Very often the indeterminate state $S = R = 1$ of an S-R flip-flop cannot be permitted to occur and then an alternative circuit, known as the **J-K flip-flop**, is used. The operational difference between the S-R and J-K flip-flops lies in the response of the circuits to the input state $S = R = 1$. The truth table of a J-K flip-flop is shown by Table 4.4 and comparing this with the truth table of an S-R flip-flop (Table 4.1) makes it clear that the J-K flip-flop *always* changes state when

Table 4.4 J-K flip-flop truth table

J	K	Q	Q^+
0	0	0	0
0	0	1	1
1	0	0	1
1	0	1	1
0	1	0	0
0	1	1	0
1	1	0	1
1	1	1	0

$J = K = 1$. Thus the J-K flip-flop *toggles* each time a clock pulse occurs.

The symbol for a J-K flip-flop is given by Fig. 4.8(*a*). A clock input is shown because this type of flip-flop is usually operated synchronously. Very often, clear and preset terminals are also provided. In addition, some J-K flip-flops are *gated*, that is they are provided with AND gates inside the ic package. The AND gate symbols are drawn touching the flip-flop symbol to indicate that the gates are internally provided (Fig. 4.8(*b*)). The J input of the flip-flop will be at logical 1 only when *all* three inputs J_1, J_2 and J_2 are at logical 1. Similarly, the K input is at 1 only when $K_1 = K_2 = K_3 = 1$.

Fig. 4.8 Symbols for (*a*) a clocked J-K flip-flop, (*b*) a gated J-K flip-flop

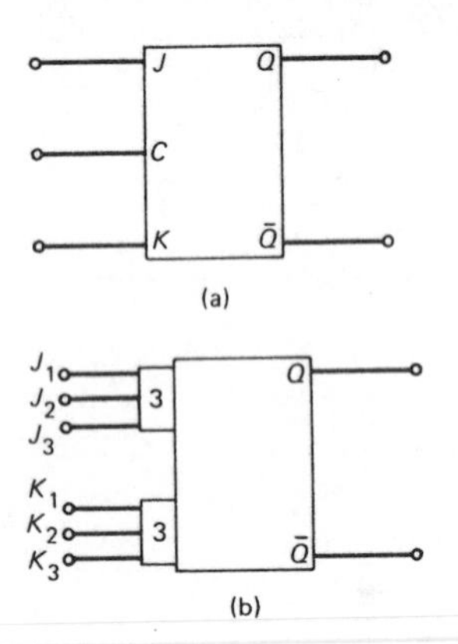

The state table of a J-K flip-flop is

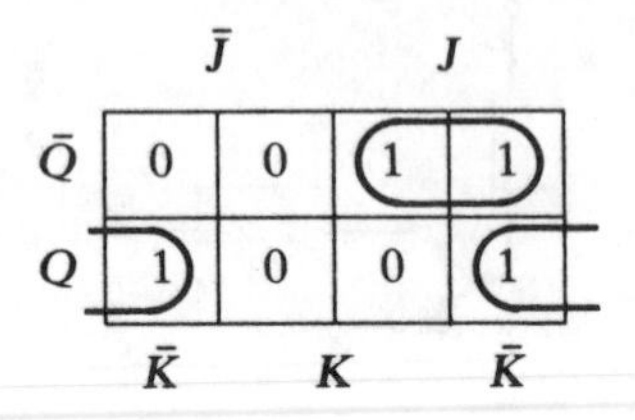

	$\bar{J}$		J	
$\bar{Q}$	0	0	1	1
Q	1	0	0	1
	$\bar{K}$	K		$\bar{K}$

and is derived from the truth table. From the state table,

$$Q^+ = J\bar{Q} + \bar{K}Q \tag{4.5}$$

$$\overline{Q^+} = \bar{J}\bar{Q} + KQ \tag{4.6}$$

The transition table of a J-K flip-flop is given by Table 4.5.

Comparing equations (4.5) and (4.6) with equations (4.2) and (4.1) it is clear that the equations differ in form only in that the J input must be ANDed with $\bar{Q}$, and the K input must be ANDed with Q. Hence the J-K flip-flop can be implemented by the circuit of Fig. 4.9(*a*). If the S-R flip-flop section of the circuit and the AND gates are implemented using NAND gates, Fig. 4.9(*b*) is obtained.

Table 4.5 J-K flip-flop transition table

Present state	*Desired next state*	*Required inputs*	
Q	Q^+	J	K
0	0	0	X
0	1	1	X
1	0	X	1
1	1	X	0

Unfortunately, in the J-K flip-flop shown in Fig. 4.9(*b*) there is a high probability of a race condition being set up when the input state is $J = K = 1$. Suppose, for example, that $Q = 0$, $\bar{Q} = 1$. When the clock changes from 0 to 1, the upper input NAND gate will be

Fig. 4.9 J-K flip-flop

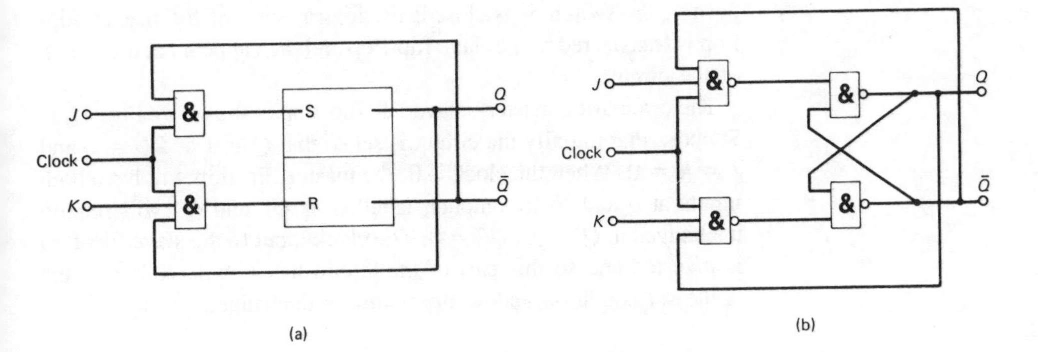

enabled and the flip-flop will set to give $Q = 1$, $\bar{Q} = 0$. Now the lower input NAND gate is enabled and the circuit resets. This gives $\bar{Q} = 1$ again and so the upper gate is enabled once more and the circuit sets again and so on for as long as the clock remains at 1. When the clock goes to 0, the circuit will remain in whichever state it happens to be in at that moment, and this is unpredictable.

Clearly, an oscillation of the output state of a flip-flop cannot be tolerated. One method of overcoming this problem is to arrange that the circuit only operates at the negative-going edge of the clock pulses; another method is to use a master–slave arrangement.

With both these types of J-K flip-flop the data present at the input terminals just prior to a clock edge determines the output Q^+ after the clock has changed state.

Master–Slave J-K Flip-Flops

The principle of operation of a **master–slave J-K flip-flop** is illustrated by the block diagram given in Fig. 4.10. The switches S_1 and S_2 are both operated by the clock. At the leading edge of the clock waveform, switch S_1 is closed and S_2 is open so that the slave flip-flop is isolated from the master flip-flop. The input data is applied

Fig. 4.10 Master–slave principle

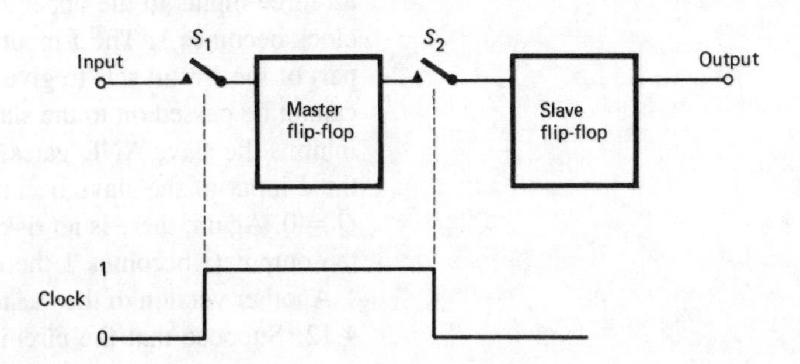

to the input (J-K) terminals of the master flip-flop and are then stored until the end of the clock pulse. When the trailing edge of the pulse occurs, the switch S_2 is closed; the logical state of the master flip-flop is transferred to the slave flip-flop and then appears at the output of the circuit.

The circuit of a master–slave J-K flip-flop is shown by Fig. 4.11. Suppose that initially the circuit is set so that $Q = 1$ and $\bar{Q} = 0$ and $J = K = 0$. When the clock is 0, the master flip-flop will have both inputs at 0 and so its outputs, labelled as Q' and $\bar{Q}'$ will remain unchanged at $Q' = 1$, $\overline{Q'} = 0$. The clock input to the slave flip-flop is inverted and so this part of the circuit has $J = 1$, $K = 0$. This is the set condition and so the output is unchanged.

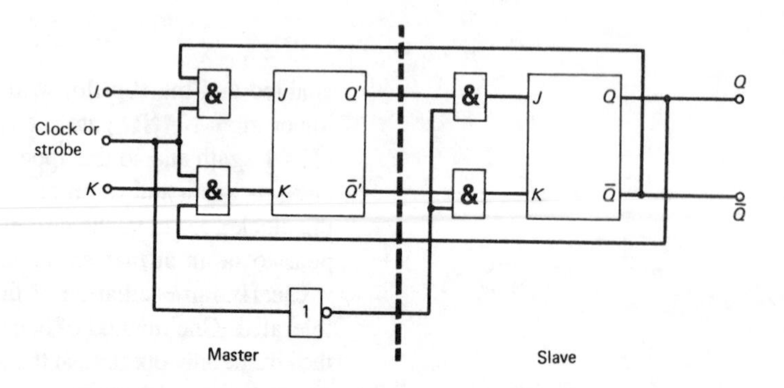

Fig. 4.11 Master-slave J-K flip-flop

When the clock is 1, there will be no change in state *unless* $J = 0$, $K = 1$ or $J = K = 1$. In either case the three inputs to the lower AND gate are all at 1 and so the master flip-flop has inputs $J = 0$, $K = 1$. The master resets to have $Q' = 0$, $\overline{Q'} = 1$. The clock input to the slave is inverted, i.e. at 0, and so both inputs to the slave remain at 0. Hence the slave does not *change* state. At the end of the clock pulse $\overline{\text{clock}}$ becomes 1 and now the slave flip-flop has $J = 0$, $K = 1$ and resets to the state $Q = 0$, $\bar{Q} = 1$. Thus the circuit does *not* change state until the end of the clock pulse. The race condition previously described cannot now occur since, when $J = \bar{Q} = 1$, the clock *must* be at 0.

Now consider the operation of the circuit when it is initially in the reset state, i.e. $Q = 0$, $\bar{Q} = 1$. When $J = 1$, $K = 0$ or $J = K = 1$, all three inputs to the upper AND gate will be at 1 immediately the clock becomes 1. The J input of the master flip-flop is 1 and so this part of the circuit sets to give $Q' = 0$, $\overline{Q'} = 0$. As before this state cannot be passed on to the slave flip-flop because the inverted clock inhibits the slave AND gates. When the clock changes from 1 to 0, the J input of the slave becomes 1 and the slave sets; now $Q = 1$, $\bar{Q} = 0$. Again, there is no risk of the circuit oscillating because when the output Q becomes 1 the clock is 0.

Another version of the master–slave J-K flip-flop is shown in Fig. 4.12. Suppose that the circuit is set with $Q = 1$, $\bar{Q} = 0$ and that

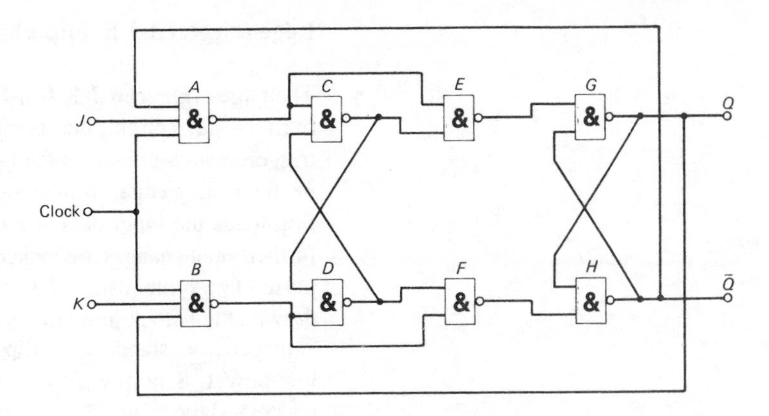

Fig. 4.12 An alternative master-slave J-K flip-flop

$J = K = 1$. When the clock is at 0, the logic states at various points in the circuit are: gates A and B both have logic 1 at their outputs and, since the master flip-flop must also be set, gate E has both its inputs at 1 and so its output is 0. Therefore, gate G has both inputs at 0 confirming that $Q = 1$. Gate F has one input at 1 and the other input at 0 so that its output is 1; thus the output of gate H is 0.

When the clock goes to 1, the output of gate B becomes 0. The two inputs to gate D are now at 1 and 0 so that its output goes to 1. This means that gate C has both inputs at 1 and its output switches to 0. The master flip-flop is now reset. Both gates E and F have one input at 1 and one input at 0 so that their outputs are 1. Now gate G has one input at 1 and the other at 0 so that its output Q *remains* at 1. Similarly, gate H has both inputs at 1 and so $\bar{Q}$ remains at 0. This means that the logic state of the master is *not* passed on to the slave.

When the clock pulse ends, the output of gate B goes to logical 1 and then gate D has inputs at 0 and 1 and so its output remains unchanged at 1. This means that gate F now has both of its inputs at 1 and so its output becomes 0. Now gate H has inputs at 0 and 1 and its output Q becomes 1. In turn, this makes both of the inputs of gate G become logical 1 and so the output Q of this gate becomes 0. The slave has now reset.

The operation of the circuit for the conditions $Q = 0$, $\bar{Q} = 1$ and $J = 1$, $K = 0$ and for $J = K = 1$ is very similar.

For any master–slave flip-flop the J and K inputs must be held constant while the clock is at logical 1, otherwise an undesirable effect known as *ones and zeros catching* will occur. If, for example, the circuit is set, and while the clock is still high the K input becomes high, the master flip-flop will reset. As a result the slave flip-flop's Q output will *catch* a 0 at the trailing edge of the clock waveform.

The problem can be overcome by the use of *data lockout* or by the use of an *edge-triggered* flip-flop. Data lockout means that data is stored in the master flip-flop during the leading-edge of the clock waveform and is then transferred to the slave flip-flop on the trailing edge.

Edge-triggered J-K Flip-Flops

The **edge-triggered J-K flip-flop** behaves in a similar manner to the master–slave circuit but its internal circuitry is different. The edge-triggered circuit reacts to the J-K inputs *only* at *either* the leading edge *or* the trailing edge, of the clock waveform. The edge-triggered flip-flop loads the input data at a clock edge but then the effects of any further input changes are locked out until the next corresponding clock pulse. The symbol for a J-K flip-flop is marked with a small wedge drawn at the clock input to indicate that the circuit is edge-triggered.

In general, standard ttl flip-flops are master–slave devices, and low-power Schottky flip-flops are edge-triggered, e.g. 7473 is master–slave, 74LS73 is edge-triggered, 7476 is master–slave, 74LS76 is edge-triggered.

The input data must be held constant for some time both before and after the clock transition takes place to ensure that it will be transferred to the output. The *set-up time* is the time that elapses between the leading edge of an input data pulse and the triggering edge of the clock pulse. Hence it is the time for which the input data must be held before the triggering edge of the clock arrives. Set-up time is illustrated for leading-edge and trailing-edge flip-flops in Fig. 4.13.

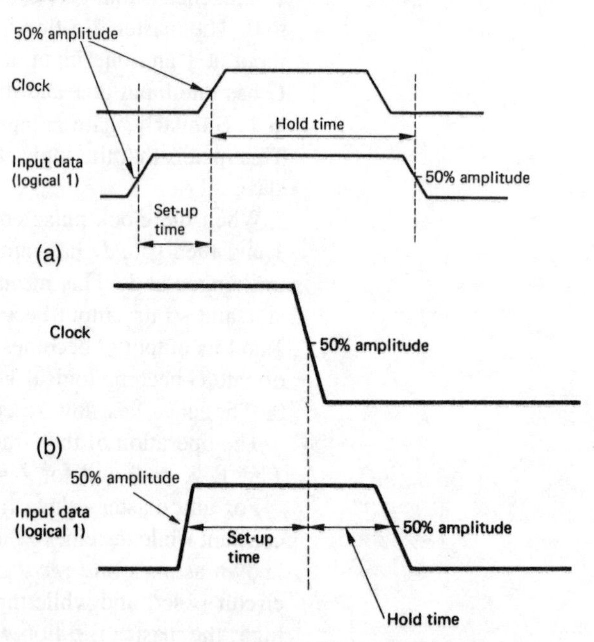

Fig. 4.13 Set-up time and hold-up time for an edge-triggered flip-flop (a) leading edge, (b) trailing edge

The *hold-up time* of an edge-triggered flip-flop is the interval of time between the clock pulse transition changing the state of the output and the end of the input data pulse. The hold time is illustrated by Fig. 4.13. After the hold time interval has passed the *J* and *K* input data may be changed without affecting the output.

Typical figures for a low-power Schottky circuit are a set-up time of 20 ns and a hold time of very nearly zero.

Clear and Preset Controls

The operation of a flip-flop is determined by the states of its input terminals, S-R or J-K, when a clock pulse is applied. The *initial* state of the circuit is not, however, so determined and may have either of its two possible values.

If Clear and Preset controls are added they can be employed to specify the initial state of the circuit. The operation of these controls is shown by the truth table of a J-K flip-flop given in Table 4.6.

Table 4.6

Preset	*Clear*	*Clock*	*J*	*K*	*Q*	Q^+
0	1	X	X	X	0	1
0	1	X	X	X	1	1
1	0	X	X	X	0	0
1	0	X	X	X	1	0
1	1	← as for J-K →				

It should be noted that the Preset and Clear terminals when at logical 0 set or reset the output Q, *whatever* the states of the clock or the J or K inputs. If both the Preset and Clear inputs are at logical 1, the operation of the circuit will be determined in a manner already described. If the Preset and/or Clear facilities are not required, the terminals must be connected to a high voltage level to disable their action.

The J-K flip-flop can be used as a *divide-by two* circuit by connecting it in the manner shown in Fig. 4.14. With the J and K input terminals connected to $+V$, both inputs are held in the logical 1 state and the circuit *toggles* with each clock pulse. The toggle action may, of course, occur at either the leading or the trailing edge of the clock pulse, depending upon the device concerned.

Fig. 4.14 The J-K flip-flop as a divide-by-two circuit

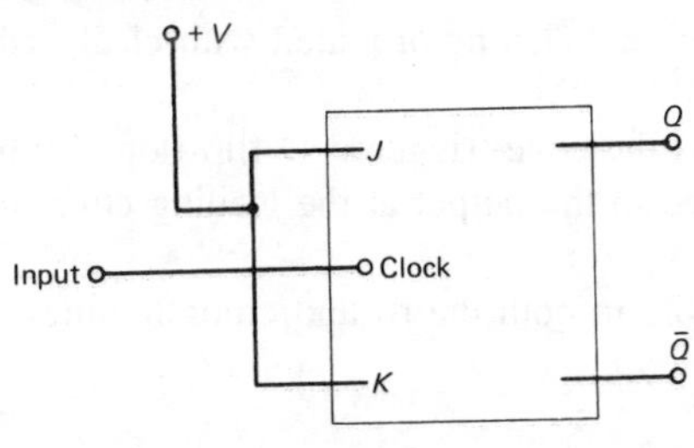

The relative merits of the two kinds of J-K flip-flops are as follows. For many master–slave circuits the J and K inputs must not be allowed to change while the clock is at logical 1 because of the effect known as "ones catching". The problem can always be overcome either by making sure that both the J and K inputs remain constant while the clock is 1, or by using short clock pulses. However, it is probably better to use edge-triggered devices instead. The data inputs to an edge-triggered device can be changed when the clock is at either 1 or 0. The new output state will be determined by the states of the J and K inputs at the start of the set-up time (during which they must, of course, be kept constant). Because the input data is not first stored by a master and then transferred to a slave, the edge-triggered flip-flop is faster to operate.

The D Flip-Flop

The **D flip-flop** has a single D input terminal. Its logical operation is such that its Q output terminal *always* takes up the same logic state as its D input. The truth table of a D flip-flop is given by Table 4.7.

Table 4.7 D flip-flop truth table

Clock	*D*	*Q*	Q^+
1	0	0	0
1	0	1	0
1	1	0	1
1	1	1	1

Any change in state occurs when the clock is 1. When the clock is 0 the output is unable to follow the D input and the circuit is said to be *latched*.

The state table of a D flip-flop is

	$\bar{D}$	D
$\bar{Q}$	0	1
Q	0	1

From the state table the characteristic equation of the circuit can be written as

$$Q^+ = D \tag{4.7}$$

$$\overline{Q^+} = \bar{D} \tag{4.8}$$

Table 4.8 D flip-flop transition table

Present state	*Desired next state*	*Required input*
Q	Q^+	D
0	0	0
0	1	1
1	0	0
1	1	1

Table 4.8 gives the transition table of the D flip-flop. Comparing it with the characteristic equation and the transition table of the J-K, or the S-R flip-flop (equations 4.1, 4.2, 4.5, and 4.6), it can be seen that the D flip-flop can be produced by a J-K or S-R flip-flop having complementary inputs. Complementary input signals can easily be arranged by connecting an inverter between the D input and the K terminal as shown by Fig. 4.15. There will not be an indeterminate state if an S-R flip-flop is used since the condition $S = R = 1$ is not possible.

The main use for the D flip-flop is as a delay circuit; the output takes up the same logical state as the input after a time delay equal to the period of one clock pulse. Some D flip-flops are manufactured in master–slave versions and may also be provided with clear and preset controls.

A variation of the circuit is the edge-triggered D flip-flop. With this circuit, data is transferred to the output at the leading edge of a clock pulse.

All D flip-flops are available in both the ttl and cmos families.

Fig. 4.15 D flip-flop

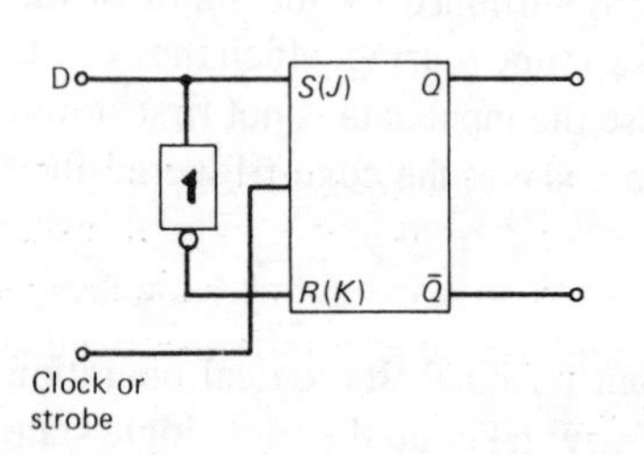

Fig. 4.16 Use of a D flip-flop to divide-by-two

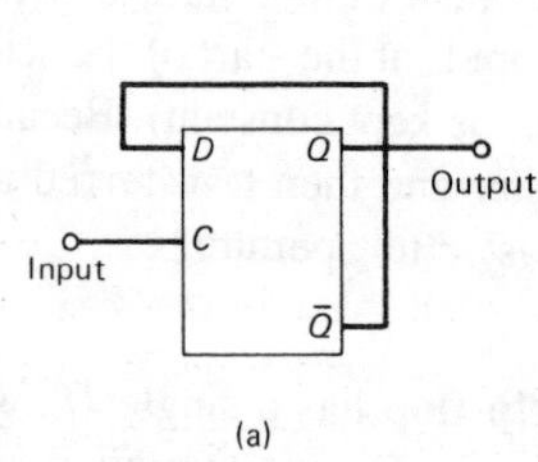

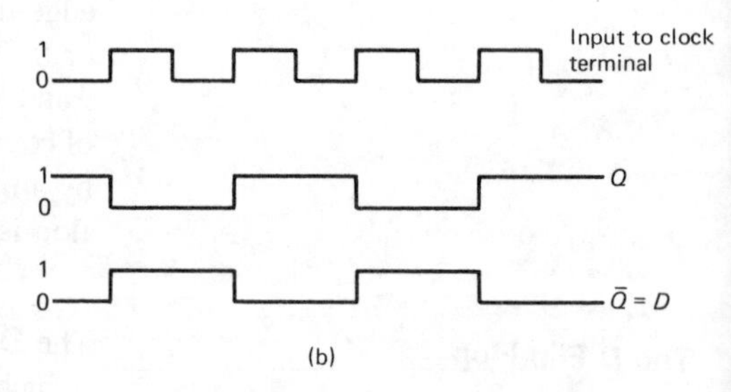

A D flip-flop can be used as a divide-by-two circuit by merely connecting its $\bar{Q}$ and D terminals together (Fig. 4.16).

The block diagram of a ttl D flip-flop is given by Fig. 4.17 (minus the clear and preset controls). Analysis of the circuit operation will show that the output Q changes to the state of the D input *only* at the leading edge of the clock pulse.

Fig. 4.17 Construction of a D flip-flop

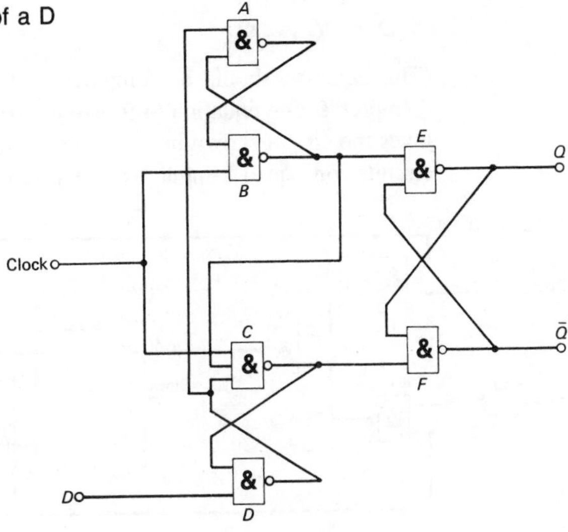

Table 4.9

Present state Q	*Desired next state* Q^+	*Required inputs* J	K	D
0	0	0	0	0
0	0	0	1	0
0	1	1	0	1
0	1	1	1	1
1	1	0	0	1
1	0	0	1	0
1	1	1	0	1
1	0	1	1	0

It has already been shown that a D flip-flop can be made from a J-K flip-flop but the reverse process is also possible. The transition table of such a circuit is given by Table 4.9. From this table the state table can be written down, mapping this time for D instead of for Q^+ (although really the same thing). Thus

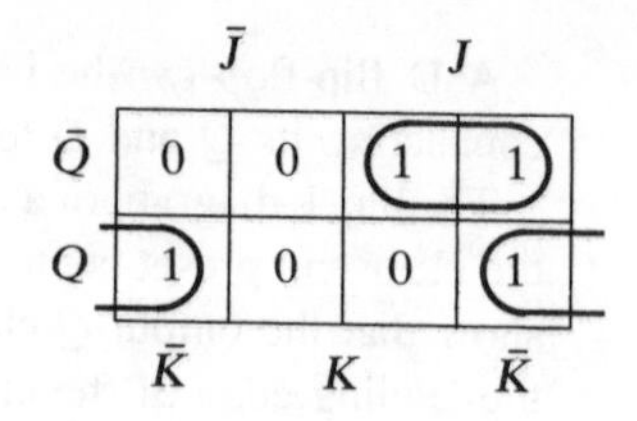

From the state table,

$$D = J\bar{Q} + \bar{K}Q \tag{4.9}$$

(This equation should be compared with equation 4.5.)

Implementing equation (4.9) using NAND gates and a D flip-flop gives the circuit shown in Fig. 4.18; it requires only two integrated circuits: one quad 2-input NAND gate and one D flip-flop.

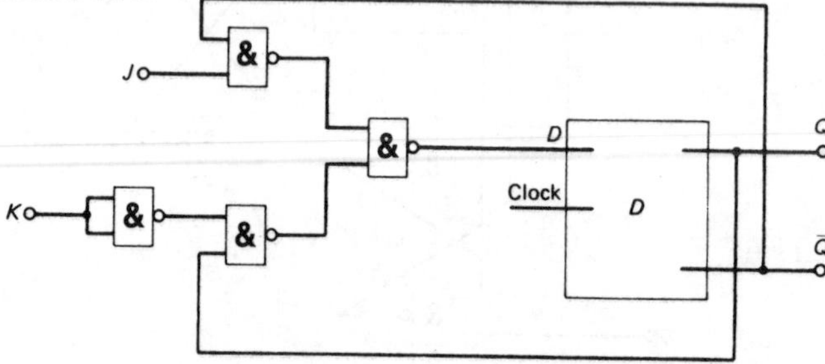

Fig. 4.18 D flip-flop connected as a J-K flip-flop

The T Flip-Flop

The fourth type of flip-flop is known as the **T flip-flop** or *trigger* flip-flop. Its truth table is given in Table 4.10. It can be seen that when the clock is 1 the Q output of the circuit will change state, or toggle, each time a pulse is applied to the trigger (T) terminal. The transition table of a T flip-flop is given by Table 4.11. Note that the columns of Tables 4.10 and 4.11 are the same. The state table is

	$\bar{T}$	T
$\bar{Q}$	0	1
Q	1	0

and from this the characteristic equations of the T flip-flop can be written as

$$Q^+ = T\bar{Q} + \bar{T}Q \tag{4.10}$$

$$\overline{Q^+} = \bar{T}\bar{Q} + TQ \tag{4.11}$$

Comparing these equations with those for the J-K flip-flop (4.5 and 4.6), it is clear that the J-K circuit can be made to act as a T flip-flop by connecting its J and K terminals together, so that at all times $J = K$ and using the common point as the T input (see Fig. 4.19(*a*)).

Fig. 4.19 T flip-flop

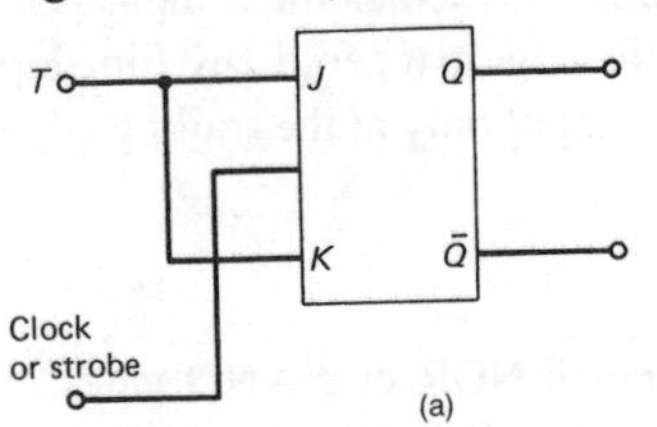

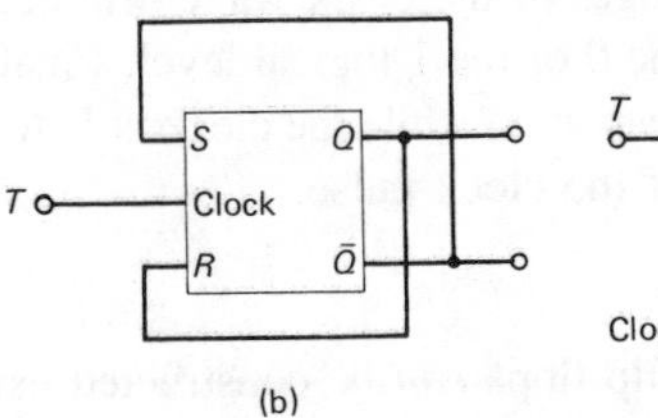

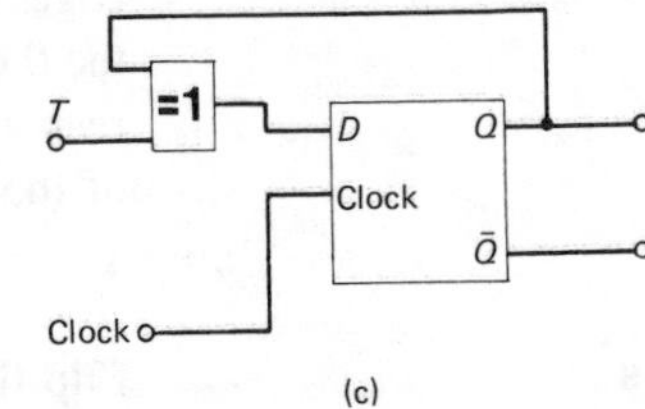

Table 4.10 T flip-flop truth table

Clock	T	Q	Q^+
1	0	0	1
1	0	1	1
1	1	0	1
1	1	1	0

Table 4.11 T flip-flop transition table

Present state	Desired next state	Required input
Q	Q^+	T
0	0	0
0	1	1
1	0	1
1	1	0

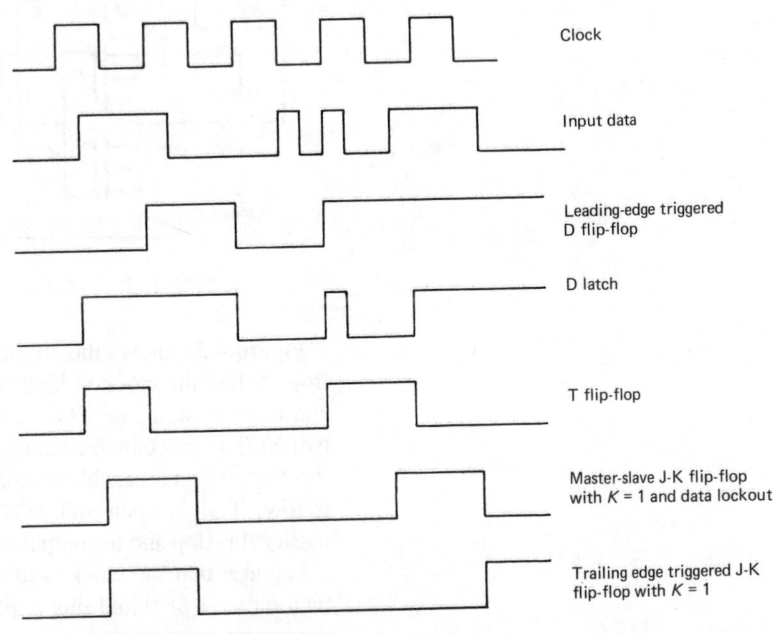

Fig. 4.20 Showing the operational difference between various types of flip-flop

The S-R and D flip-flops can also be connected as a T flip-flop (see Fig. 4.19(*b*) and (*c*)).

The *differences between the various types of flip-flop* can be seen with the aid of a timing diagram (Fig. 4.20). The D flip-flop changes its state to be the same as the input state only at the leading edge of the clock waveform. The D latch has an output that is the same as the input whenever the clock is at the logical 1 voltage level. The T flip-flop changes its state whenever both the clock and the input are at 1. The master–slave J-K flip-flop changes its state at the trailing edges of the clock waveform according to whether the *J* input is at the 0 or the 1 logical level. Finally, the edge-triggered J-K flip-flop responds, while the clock is 1, to the *J* input only at the trailing edge of the clock pulse.

CMOS Flip-Flops

Flip-flops *can* be constructed using cmos NOR or NAND gates in the ways previously discussed but cmos flip-flops do not use this approach. This is because transmission gates (p. 59) are easy to make using cmos techniques and their use allows different, and simpler, circuitry to be used.

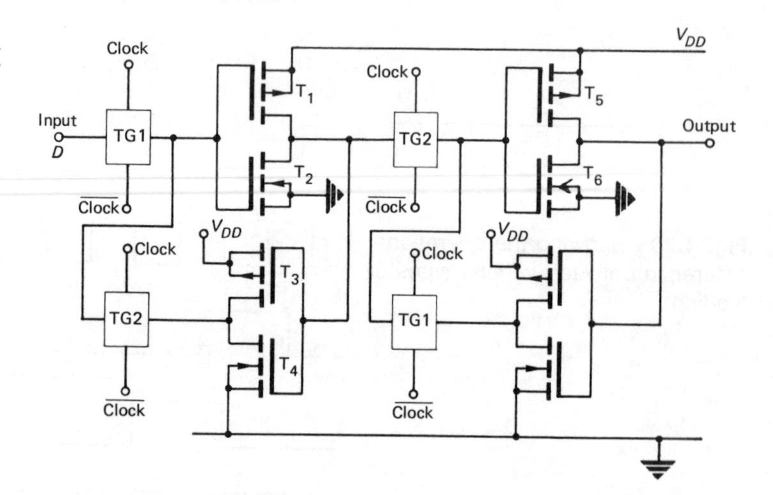

Fig. 4.21 CMOS master–slave J-K flip-flop

Figure 4.21 shows the circuit of a **cmos master–slave J-K flip-flop**. When the clock is high, the transmission gate TG1 is closed and transmission gate TG2 is open. The mosfets T_5–T_8 then form two NOT gates connected in cascade and thus provide a simple slave flip-flop, which is capable of storing one bit. Similarly, when the clock is low, TG1 is open and TG2 is closed; now T_1–T_4 provide the master flip-flop and the output of the master is connected to the slave.

Suppose that the clock is at logical 0. The state of the master is then either 1 or 0 and this is passed on to the slave and appears at the output terminals of the circuit. The master is isolated from the input. If, now, the clock goes to 1, gate TG1 closes and the state of the input is sampled. This state is not passed on to the slave because TG2 is now open. The sampled data is held by the master, until the end of the clock pulse, by the flip-flop formed by $T_1 - T_4$.

Flip-flops are also manufactured using the ecl and I^2L techniques, although the number of ecl devices available is very limited compared with ttl and cmos, while I^2L is exclusively an lsi technology. Their relative merits have been discussed in Chapter 2.

Exercises 4

4.1 With the aid of a diagram explain the operation of an S-R flip-flop. List the disadvantages of the S-R flip-flop and explain how these are overcome by the master–slave J-K flip-flop.

4.2 Analyse the operation of the D flip-flop given in Fig. 4.17.

4.3 Write down the truth table of an S-R flip-flop and use it to obtain *a*) the state table, *b*) the characteristic equation, *c*) the transition table of the circuit. Use your results to derive *either* the NOR *or* the NAND gate version of the circuit.

4.4 Write down the truth tables for the J-K, D and T flip-flops. Hence obtain the characteristic equation of each circuit. Compare your results and hence explain how the J-K circuit can be operated as *a*) a D flip-flop, *b*) a T flip-flop.

4.5 Briefly explain the principle of operation of master–slave and edge-triggered J-K flip-flops. Most low-power Schottky versions of J-K flip-flops are edge-triggered devices. Suggest some reasons for this.

4.6 *a*) Draw the circuit of an S-R flip-flop that uses NOR gates only. Modify the circuit to produce a D flip-flop. In both cases give the truth table of the circuit.
b) Explain how a D flip-flop can be used as a divide-by-two circuit. Give the truth table of the divider.

4.7 Draw the logic circuit of a either a NAND or a NOR gate clocked latch. By means of a truth table show how the Q and $\bar{Q}$ outputs respond to signals applied to the S and R inputs. Hence state the basic disadvantage of the latch.

4.8 *a*) Define the terms *set-up time* and *hold time* when applied to an edge-triggered flip-flop. Give typical values for a ttl device and explain the importance of these quantities.
b) Many flip-flop circuits are provided with clear and preset controls. Explain their functions.

4.9 Show how the master–slave construction of a J-K flip-flop introduces a delay between input and output and explain the reason for this arrangement. Also explain how the circuit may be made to "toggle" and sketch the output waveform for three cycles of the clock input.

5 Counters and Shift Registers

Introduction

A **counter** is a digital circuit that consists of a number n of flip-flops connected in cascade, whose function is to count the number of pulses applied to an input terminal. The count may be indicated using the straightforward binary code, or binary-coded-decimal, or some other code. Alternatively, decoding circuitry may be employed to give a direct readout of the count. If only the Q output of the final flip-flop is employed the counter will act as a divide-by-n circuit. The maximum number of possible 1 and 0 states is known as the *modulus* of the counter and this cannot be greater than 2^n.

A counter can be constructed using a number of ic flip-flop packages, with some gating if the desired count is less than 2^n, or a number of msi counters are available in both the ttl and the cmos facilities. In either case the counter may be operated either synchronously or non-synchronously. Most ic counters are either 4-bit ($\div$ 16) or decade counters which produce a straight binary output that can, if required, have their count reduced. Counters must be individually designed, if an irregular sequence, such as a unit-distance code, is required.

Counters find application in many kinds of digital equipments, such as for example, the direct counting of pulses, division-by-n, and frequency and time measurements.

A **shift register** also consists of a number of flip-flops and operates in a similar manner to a counter. Shift registers are used as temporary stores, serial-to-parallel converters (and vice versa), and can also be connected to operate as *ring counters* and sequence generators.

Non-synchronous Counters

A single J-K or D flip-flop can be connected to act as a divide-by-two circuit (Figs. 4.14 and 4.16). For counts greater than 2, a number n of flip-flops must be connected in cascade to give a maximum count of $2^n - 1$. Since the first count is 0 this means that n flip-flops can give 2^n binary states.

The idea is illustrated by Figs. 5.1(*a*), (*b*) and (*c*). Figure 5.1(*a*) shows how two D flip-flops can be connected to give a maximum count of 3 (divide-by-4) and Fig. 5.1(*b*) shows three J-K flip-flops connected as a divide-by-8 circuit. The J and K inputs of each flip-flop are connected to the logical 1 voltage level. Throughout this

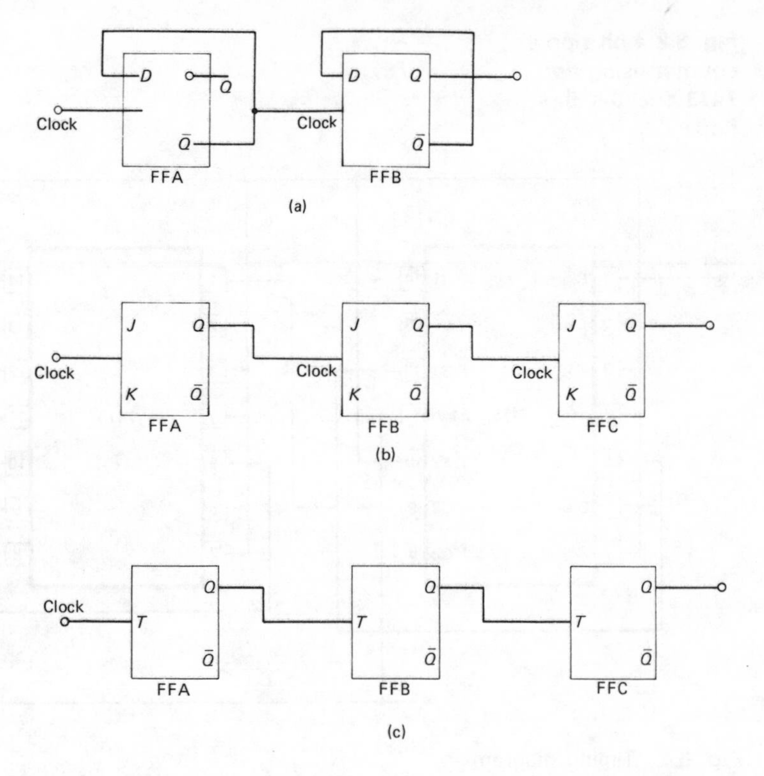

Fig. 5.1 (*a*) D flip-flops connected as a divide-by-4 circuit (*b*) J-K flip-flops connected as a divide-by-8 circuit (*c*) T flip-flops connected as a divide-by-8 circuit

chapter, if $J = K = 1$, it will not be shown. Each stage is clocked at one-half the rate of the preceding stage and so the states of the Q outputs appear to *ripple* through the circuit. For this reason the circuits are often known as *ripple counters*. Figure 5.1(*c*) shows three T flip-flops connected as a ripple counter.

The main disadvantages of this type of circuit are that the propagation delays of the individual stages are additive and the total delay may severely limit the maximum possible frequency of operation. Also, the count can only follow a straight binary code sequence.

Example 5.1

The flip-flops in a 4-bit counter each introduce a maximum delay of 40 ns. Calculate the maximum clock frequency.

Solution

Total propagation delay $= 4 \times 40 = 160$ ns.

Thus, maximum clock frequency $= 1/(160 \times 10^{-9}) = 6.25 \times 10^{6}$ (*Ans*)

Four-stage (bit) ripple counters can also be designed using an extension of the method shown in Fig. 5.1. Figure 5.2 shows a *4-bit counter* that uses two 7473 dual J-K master–slave flip-flops. The timing diagram of the 4-bit counter is given in Fig. 5.3 and its truth table

Fig. 5.2 4-bit ripple counter using two 7473 dual J-K flip-flops

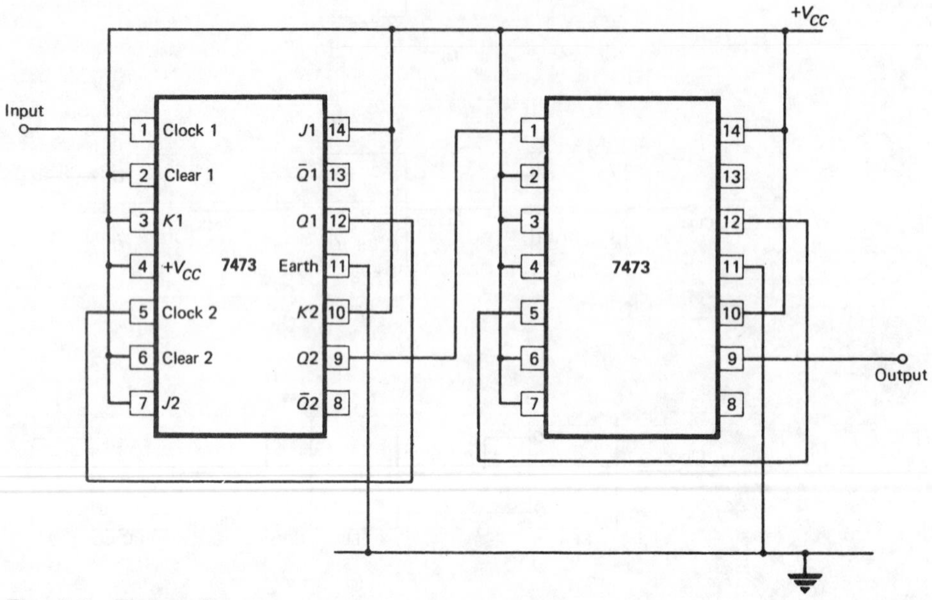

Fig. 5.3 Timing diagram of a 4-bit ripple counter

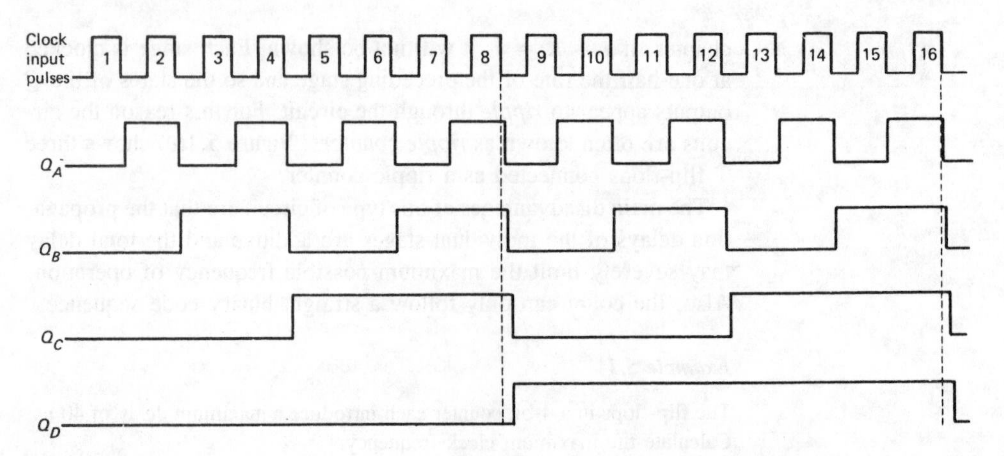

Table 5.1 4-bit counter truth table

Clock pulse	0	1	2	3	4	5	6	7	8	9	10	11	12	13	14	15	16
Q_A	0	1	0	1	0	1	0	1	0	1	0	1	0	1	0	1	0
Q_B	0	0	1	1	0	0	1	1	0	0	1	1	0	0	1	1	0
Q_C	0	0	0	0	1	1	1	1	0	0	0	0	1	1	1	1	0
Q_D	0	0	0	0	0	0	0	0	1	1	1	1	1	1	1	1	0

is given by Table 5.1 and these can be used to explain the action of the circuit.

Suppose that, initially, the circuit has each of its four stages reset, i.e. $Q_A = Q_B = Q_C = Q_D = 0$. The trailing edge of the first clock pulse will cause the first flip-flop to toggle so that $Q_A = 1$. The second clock pulse will again toggle flip-flop 1 and the change of Q_A from 1 to 0 will toggle the second flip-flop. Now $Q_A = 0$, $Q_B = 1$. Clock pulse 3 will toggle the first flip-flop so that Q_A becomes 1 but the change in Q_A from 0 to 1 will have no effect on the logical state of flip-flop 2. Now $Q_A = Q_B = 1$ and $Q_C = Q_D = 0$. The fourth clock pulse causes Q_A to change from 1 to 0 and this change in state resets the second flip-flop. Q_B goes from 1 to 0 and in so doing causes the third flip-flop to toggle. Now, $Q_A = Q_B = Q_D = 0$ and $Q_C = 1$.

Table 5.2

Clock pulse	0	1	2	3	4	5	6	7	8	9	10	11	12	13	14	15	16
$\bar{Q}_A$	1	0	1	0	1	0	1	0	1	0	1	0	1	0	1	0	1
$\bar{Q}_B$	1	1	0	0	1	1	0	0	1	1	0	0	1	1	0	0	1
$\bar{Q}_C$	1	1	1	1	0	0	0	0	1	1	1	1	0	0	0	0	1
$\bar{Q}_D$	1	1	1	1	1	1	1	1	0	0	0	0	0	0	0	0	1

Further operation of the counter as succeeding clock pulses are applied to the circuit is summarized by its truth table and the timing diagram. If the $\bar{Q}$ outputs are employed to give a binary readout, the circuit will count down from 15 to 0. The truth table describing this action is given by Table 5.2. It should be compared with Table 5.1.

A ripple counter will also count down (instead of up) if the $\bar{Q}$ outputs of each stage are used as the clock signal for the following stage and the Q outputs are employed.

When more than two or, perhaps, three stages are required it is usually cheaper to use an msi counter. Examples of these include: 7493 4-bit counter, 74293 4-bit counter, 4024 7-bit counter, 4040 12-bit counter, and 4020 14-bit counter.

If the output is taken from the Q terminal of the final flip-flop, as in Figs. 5.1 and 5.2, the circuit acts as divider. The count at any time can be indicated by utilizing the Q outputs of each stage and some kind of indicating device, e.g. LEDs.

For some applications a binary readout is undesirable; perhaps the circuit is required to generate an output pulse when the count reaches some particular number. When this is the case a decoder must be added to the circuit. Decoding is easily achieved by connecting the Q output of each stage that is at 1, and the $\bar{Q}$ output of each stage that is at 0, for the required count to a NAND gate. The output of the NAND gate will be at logical 0 *only* when the count is at the required point.

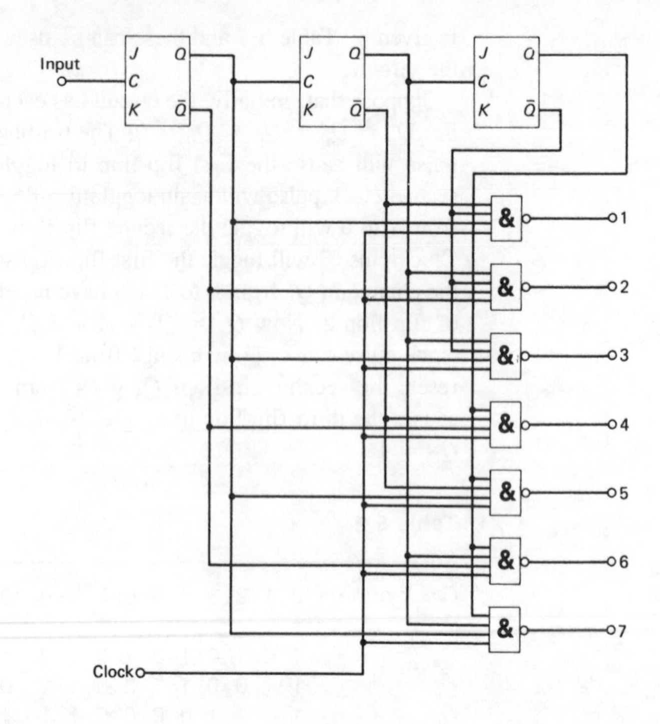

Fig. 5.4 Counter with decoded outputs

An example of the technique is shown in Fig. 5.4. The enabling clock pulse is not applied until all the flip-flops have switched and this ensures that false signals, or glitches, do not appear at the output.

Reducing the Count to Less than 2^n

Very often a counter is required to have a count of less than 2^n, where n is the number of stages. A ripple counter must then be modified so that one or more of the possible counts are omitted. For example, if a count of 6 is required, a 3-stage counter must be used with *two* of its counts eliminated. The reduction of the count is achieved by decoding the maximum count number wanted and using a signal derived from it to clear the counter.

The design of the reset type of counter must start with a decision on the number n of stages that are necessary. A NAND gate is then required whose output is connected to all the reset, or clear, flip-flop inputs in parallel. The inputs to the NAND gate must consist of the Q output of each stage that is at logical 1 when the required count is reached.

1 *Decade counter* Four stages are necessary: for a count of 10_{10} or 1010, then $Q_A = 0$, $Q_B = 1$, $Q_C = 0$, $Q_D = 1$, and hence the NAND gate inputs are Q_B and Q_D. The circuit is shown by Fig. 5.5. The timing diagram of Fig. 5.6 illustrates the operation of the decade counter.

Fig. 5.5 Non-synchronous decade counter

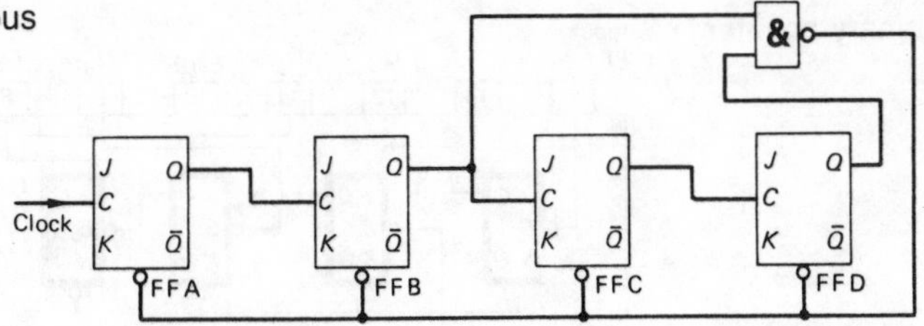

Fig. 5.6 Timing diagram of a decade counter

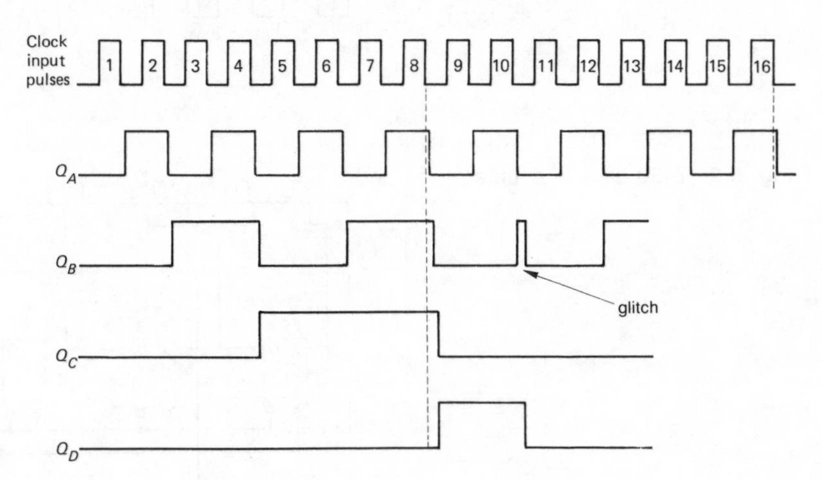

Fig. 5.7 Divide-by-6 ripple counter

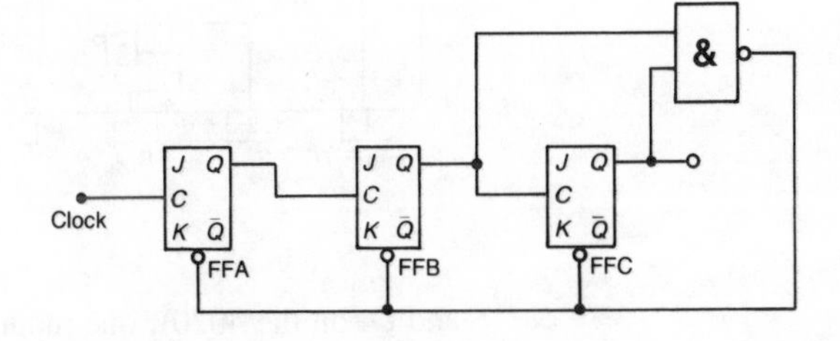

2 *Divide-by-6 counter* Three stages are needed: for a count of 6 or 110, then $Q_A = 0$, $Q_B = 1$, $Q_C = 1$ and these are the necessary NAND gate inputs (Fig. 5.7).

3 *Divide-by-5 counter* Three stages are needed: for a count of 5 or 101, then $Q_A = 1$ and $Q_C = 1$ and so the NAND gate inputs are Q_A and Q_C. This counter is easily converted into a decade counter by preceding it with a D or a J-K flip-flop connected as a divide-by-2 circuit. This is the basis of the ttl 7490 decade counter (Fig. 5.9).

Ripple counters are available in the ttl and cmos logic families. Examples are

a) Cmos: 4020 14-stage, 4024 7-stage, 4040 12-stage ripple counters.

The 4020 and 4040 circuits have buffered output terminals (not Q_2

Fig. 5.8 7493 4-bit binary counter

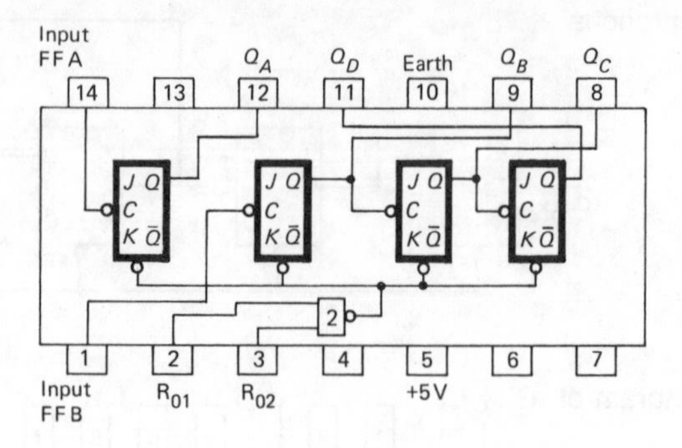

Fig. 5.9 7490 decade counter

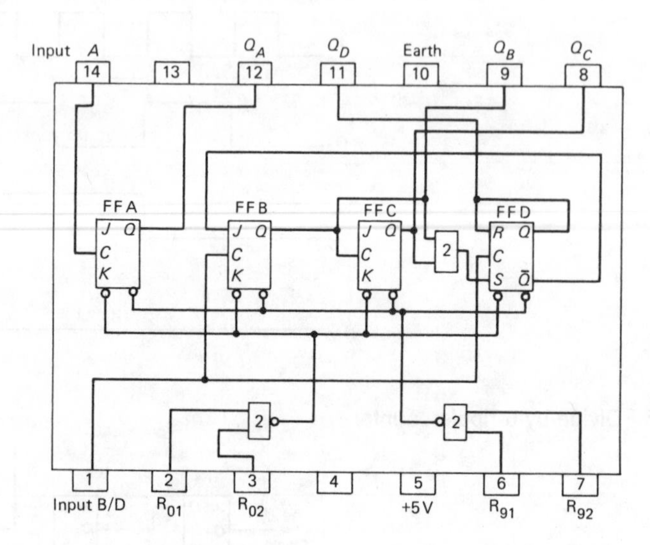

and Q_3 on the 4020), one input terminal, and a reset terminal, while the 4024 has seven buffered outputs and reset and input terminals. These devices can be made into divide-by-n counters using the reset method.

b) TTL: 7490 decade counter, 7492 divide-by-12 counter, 7493 4-bit binary ripple counter, 74196 presettable decade counter, 74197 presettable 4-bit ripple counter, and 74390 dual decade ripple counter.

All of these devices are available in both the standard and the low-power Schottky versions.

The 7493 consists of two sections: one a single divide-by-two circuit, the other three J-K flip-flops connected as a divide-by-8 ripple counter. For the circuit to act as a divide-by-16 counter the Q output of the first flip-flop must be connected to the clock input of the second flip-flop. Referring to Fig. 5.8, terminals 1 and 12 are linked together.

The 7493 can be modified, using the reset method, to give a count other than 2, 8 or 16. The circuit will reset at the required count if the Q outputs that are then at 1 are fed back to the reset inputs R_{01}

Table 5.3

Divide-by-n	R_{01} to:	R_{02} to:	
$n = 7$	Q_A	$Q_B Q_C$	(via gating)
$n = 9$	Q_A	Q_D	
$n = 10$	Q_B	Q_D	
$n = 11$	Q_A	$Q_B Q_D$	(via gating)
$n = 12$	Q_C	Q_D	
$n = 13$	Q_A	$Q_C Q_D$	(via gating)
$n = 14$	$Q_B Q_D$	Q_C	(via gating)
$n = 15$	Q_A	$Q_B Q_C Q_D$	(via gating)

and R_{02}. Table 5.3 gives examples of the connections necessary for various division ratios. An external gate may sometimes be necessary.

The 7490 decade counter can also be modified to produce alternative counts. This device (Fig. 5.9) consists of four master–slave J-K flip-flops connected to provide a divide-by-2 counter and a divide-by-5 counter. It can be seen from the figure how the three flip-flops are connected to give a count of 5. In addition to the reset terminals R_{01} and R_{02}, two further terminals R_{91} and R_{92} are provided which allow the counter to be set to a count of 9. When this facility is not required these two terminals should be earthed.

For operation as a bcd decade counter, the B/D input pin must be connected to the Q_A terminal but if a symmetrical divide-by-10 count is needed then Q_D should be connected to the A input. The input signal is then applied to the B/D input terminal and the output is taken from the Q_A terminal. As for the 7493 the outputs that are high at the required count must be connected to the R_{01} and R_{02} terminals, e.g. for a count of 6 connect Q_B to R_{01} and Q_C to R_{02}. The timing diagram for this particular case is given in Fig. 5.10. Q_B goes to logic 1 at the trailing edge of the second pulse and then goes back to 0 at the end of the fourth pulse. At this instant Q_C becomes 1.

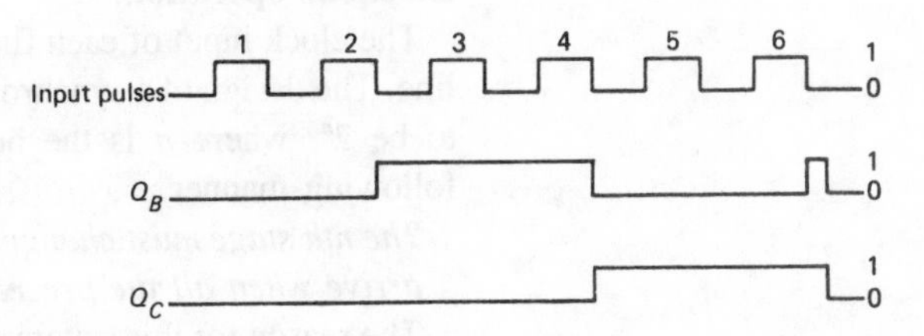

Fig. 5.10 Timing diagram for a divide-by-6 counter

When the trailing edge of the sixth input pulse occurs, Q_B goes to 1 again and now both Q_B *and* Q_C are 1 and so the circuit resets its count to 0. Thus, Q_B is only at 1 for a very short period of time so that a voltage spike appears at the output terminal.

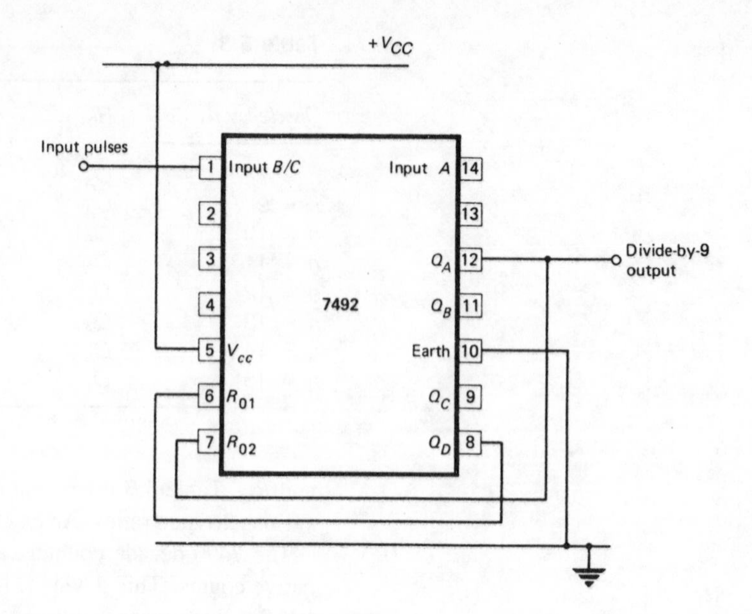

Fig. 5.11 7492 connected as a divide-by-9 counter

In similar manner the 7492 divide-by-12 counter can also be connected for counts of less than 12. Figure 5.11 shows one example, a divide-by-9 counter. Notice that once again the Q outputs that are high at the required count of 9 (1001) are connected to the reset terminals.

The simple reset method, as described for the 7490, the 7492 and the 7493, possesses the disadvantage that the reset pulse only lasts for as long as it takes for one of the "high" flip-flops to reset. This can, in some circumstances, lead to difficulties.

Synchronous Counters

For many applications of counters the time taken for a clock pulse to ripple through the circuit and/or the glitches that often arise are not acceptable. The speed of operation can be considerably increased if all the flip-flops are simultaneously clocked. This is known as **synchronous operation.**

The clock input of each flip-flop is directly connected to the clock line. The design of a synchronous counter when the count is required to be 2^n, where n is the number of stages, is determined in the following manner.

The nth stage must change its state only for those clock pulses that arrive when all the preceding stages are set, i.e. have $Q = 1$.

The reason for this statement can be understood by considering the required values for Q_A, Q_B, etc. as the count increases. For example, for a count of 5, $Q_A = 1$, $Q_B = 0$, $Q_C = 1$ and $Q_D = 0$.

Suppose that a divide-by-8 counter is to be designed. Three stages are necessary. The first stage must toggle on each clock pulse and so $J_A = K_A = 1$. The second stage must toggle only when $Q_A = 1$ and to achieve this Q_A is connected to both J_B and K_B. Stage C must

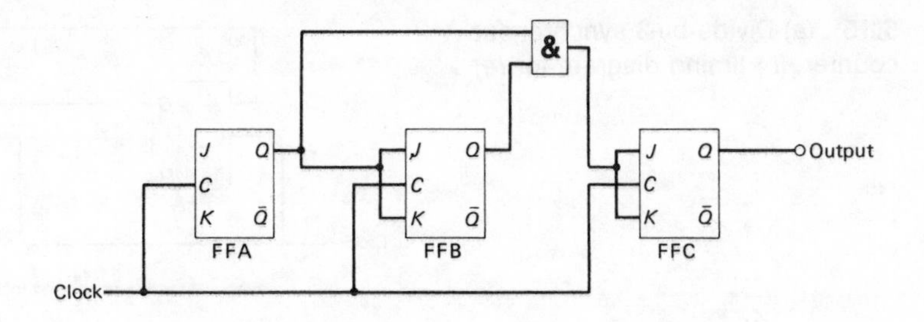

Fig. 5.12 Divide-by-8 synchronous counter

toggle only when *both* Q_A and Q_B are high and so these terminals are connected to the inputs of a 2-input AND gate. The output of the AND gate is then connected to both J_C and K_C. Thus the required circuit is shown by Fig. 5.12.

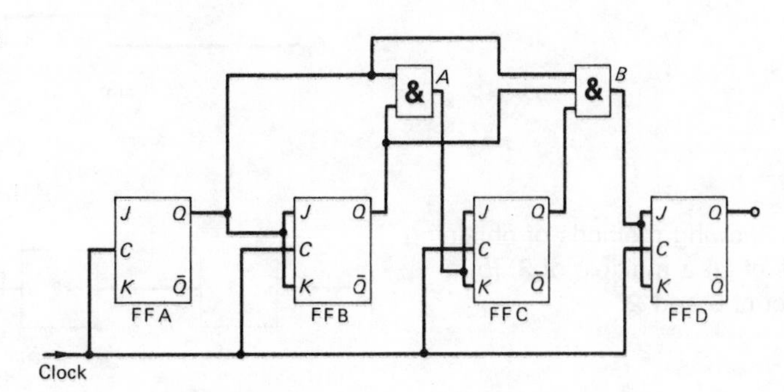

Fig. 5.13 Divide-by-16 synchronous counter

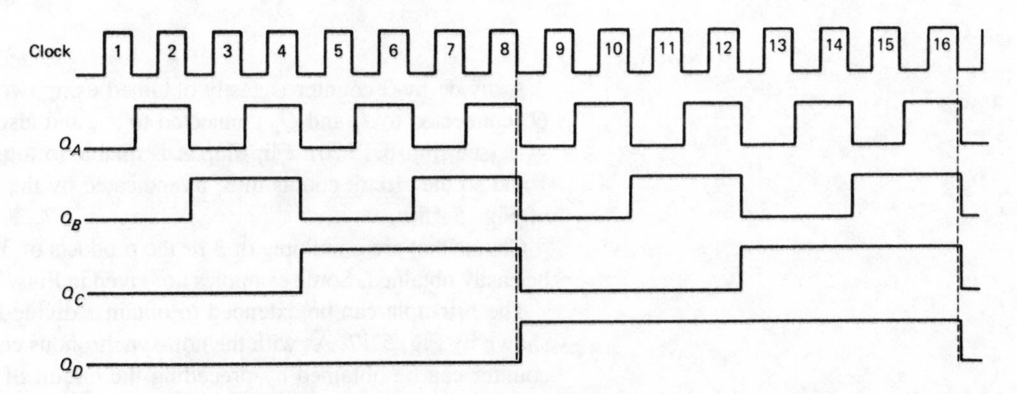

Fig. 5.14 Timing diagram for a 4-bit synchronous counter

This principle is easily extended to obtain a 4-bit (divide-by-16) counter, a 5-bit (divide-by-32) counter, and so on, and Fig. 5.13 shows the circuit of the 4-bit counter. The operation of the 4-bit synchronous counter is summarized by its timing diagram, given by Fig. 5.14.

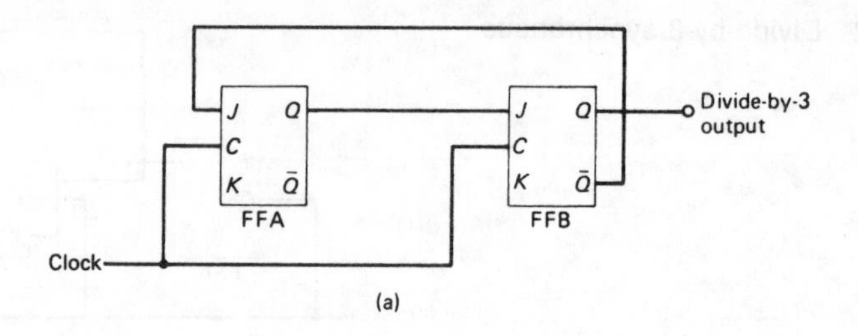

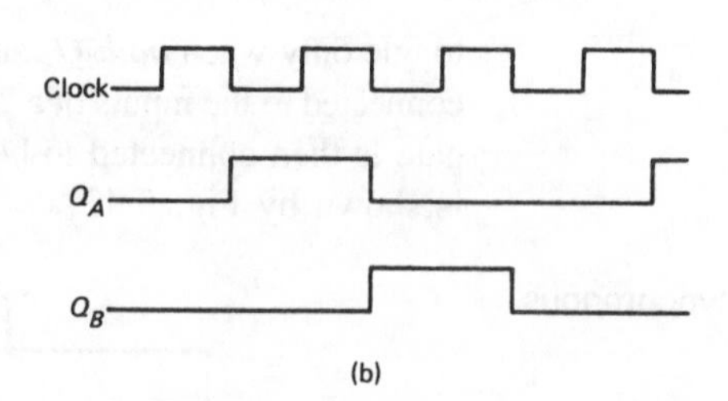

5.15 (*a*) Divide-by-3 synchronous counter, (*b*) timing diagram for (*a*)

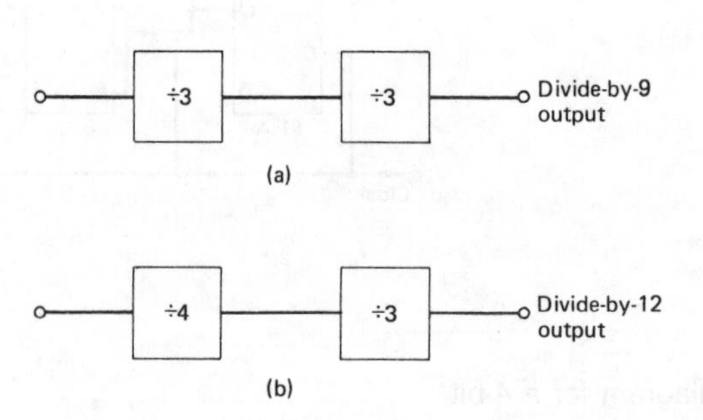

5.16 Showing methods of obtaining counts of (*a*) a multiple of 3, (*b*) a product of 3 and 2^n

A divide-by-3 counter is easily obtained using two flip-flops with Q_A connected to J_B and $\bar{Q}_B$ connected to J_A, and also with $K_A = K_B = 1$ (see Fig. 5.15(*a*)). Flip-flop *A* is unable to toggle when $Q_B = 1$ and so the circuit counts to 3, as indicated by the timing diagram of Fig. 5.15(*b*).

Counts that are a multiple of 3 *or* the products of 3 and 2^n can also be easily obtained. Some examples are given in Figs. 5.16(*a*) and (*b*).

The principle can be extended to obtain a divide-by-5 counter as shown by Fig. 5.17. As with the non-synchronous counter, a decade counter can be obtained by preceding the circuit of Fig. 5.17 by a divide-by-2 circuit.

For other counts, and when an irregular count sequence is wanted, a more complex design method is necessary. The aim of the design method is to produce the minimal Boolean expressions for the gating required for the *D* or the *J* and *K* inputs of each flip-flop in the counter.

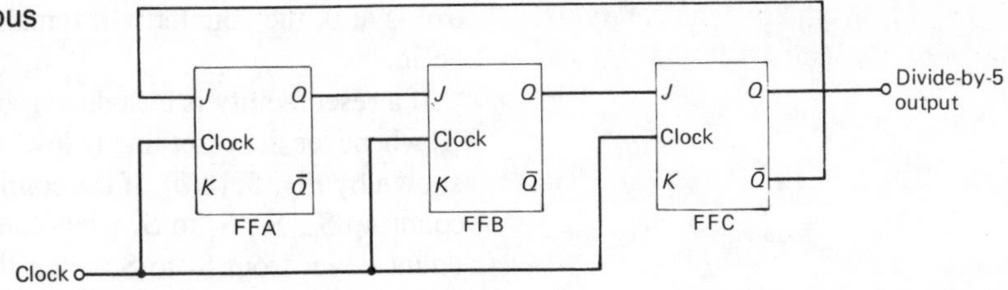

Fig. 5.17 Divide-by-5 synchronous counter

State Diagrams

The operation of a synchronous counter can be described by means of a **state diagram** which relates the present state of the Q outputs, and the next Q^+ states, of each flip-flop to the inputs and to the present output states. The diagram should include all the possible state transitions. Each of the possible states of the counter is represented by a circle labelled as S_1, S_2, S_3, etc. Possible transitions between these states are indicated by lines drawn to link the appropriate circles. Each line is labelled with two binary numbers separated by /. The left-hand number indicates the input that determines the next count state and the right-hand number indicates the output state that is then entered.

Consider a 2-bit binary counter. This will have just four possible states; these are 00, 01, 10, 11 and they will be labelled respectively as S_0, S_1, S_2 and S_3. The state diagram for this counter is given by Fig. 5.18(*a*). As long as the control is at 1, the count proceeds sequentially from state S_0 to state S_3 and then back again to S_0. If the con-

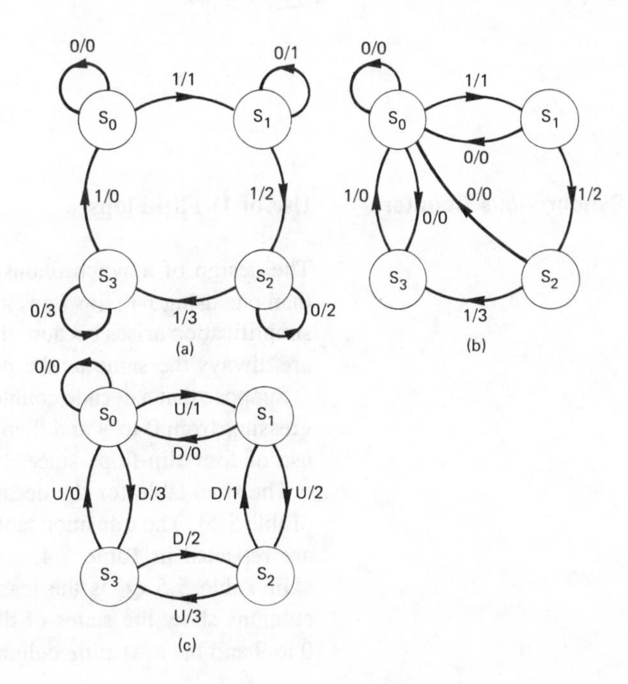

Fig. 5.18 State diagrams for (*a*) 2-bit counter, (*b*) resettable 2-bit counter, (*c*) up-down 2-bit counter

trol is at 0, the counter will remain in whatever state it happened to be in.

If a reset facility is introduced so that the counter will reset to state S_0 whenever the reset line is low, the state diagram is altered to that shown by Fig. 5.18(*b*). If the counter is of the up–down type, it will count up S_0, S_1, S_2 to S_3 when the control up-line is high, and will count down from S_3 to S_0 when the down control is high (see Fig. 5.18(*c*)).

Some other state diagrams are shown in Fig. 5.19: (*a*) represents a divide-by-7 counter, (*b*) is a divide-by-7 up–down counter, (*c*) is a divide-by-5 resettable counter, and (*d*) is a decade counter.

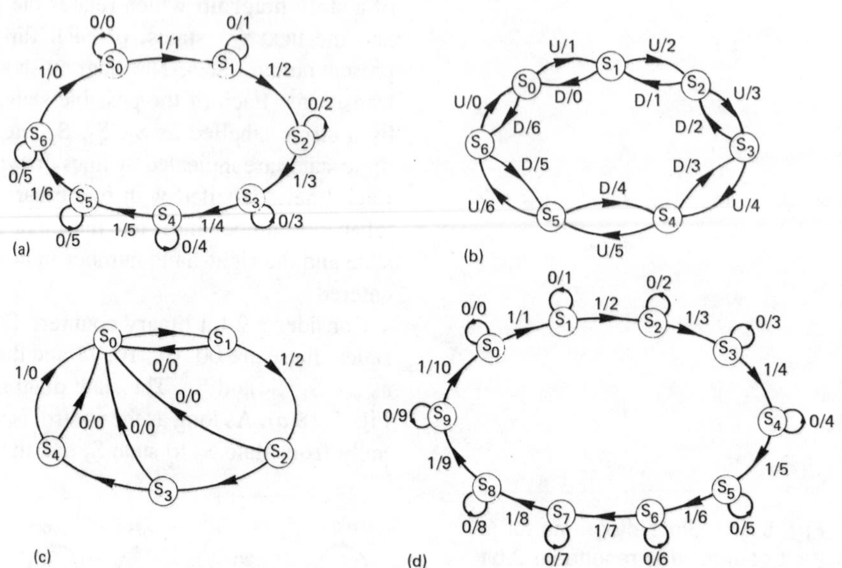

Fig. 5.19 State diagrams for (*a*) divide-by-7 counter, (*b*) divide-by-7 up-down counter, (*c*) divide-by-5 resettable counter, and (*d*) decade counter

Design of Synchronous Counters

Use of D Flip-Flops

The design of a synchronous counter using D flip-flops is simpler than one using J-K flip-flops although the basic idea is the same. The simplification arises because the next-state outputs of each D flip-flop are always the same as the present-state D inputs.

Suppose that a decade counter is to be designed with the count progressing from 0 to 9 and then back to 0. A count of 10 requires the use of four flip-flops since $2^3 = 8$.

The state table for the decade counter must first be written down (Table 5.5). The transition tables of D and J-K flip-flops given earlier are repeated in Table 5.4.

In Table 5.5 Q_A is the least significant output; the *present state* columns show the states of the four flip-flops for each count from 0 to 9 and the *next state* columns show the required changes, if any,

Table 5.4 D and J-K flip-flop transition table

Q	Q^+	J	K	D
0	0	0	X	0
0	1	1	X	1
1	0	X	1	0
1	1	X	0	1

Table 5.5 Decade counter state table (D flip-flops)

	Present state								Next state			
	Q_D	Q_C	Q_B	Q_A	D_D	D_C	D_B	D_A	Q_D^+	Q_B^+	Q_C^+	Q_A^+
0	0	0	0	0	0	0	0	1	0	0	0	1
1	0	0	0	1	0	0	1	0	0	0	1	0
2	0	0	1	0	0	0	1	1	0	0	1	1
3	0	0	1	1	0	1	0	0	0	1	0	0
4	0	1	0	0	0	1	0	1	0	1	0	1
5	0	1	0	1	0	1	1	0	0	1	1	0
6	0	1	1	0	0	1	1	1	0	1	1	1
7	0	1	1	1	1	0	0	0	1	0	0	0
8	1	0	0	0	1	0	0	1	1	0	0	1
9	1	0	0	1	1	0	1	0	1	0	1	0
10	1	0	1	0	X	X	X	X	X	X	X	X
↓												
15	X	X	X	X	X	X	X	X	X	X	X	X

in Q_A, Q_B, Q_C and Q_D. For counts 10–15, the states of Q_A, etc. are don't-cares and are shown by the symbol X. Using the D flip-flop transition table (Table 5.4) the required D inputs can be listed. Note that when the next state is a don't care the D inputs are also don't cares.

The next step is to draw a state mapping for each of the D inputs.

	Q_A		$\overline{Q_A}$		
Q_C	X	X	X	X	Q_D
Q_C	0	0	1	1	$\overline{Q_D}$
$\overline{Q_C}$	0	0	1	1	$\overline{Q_D}$
$\overline{Q_C}$	X	0	1	X	Q_D
	Q_B	$\overline{Q_B}$		Q_B	

1 $D_A = \overline{Q_A}$

	Q_A		$\overline{Q_A}$		
Q_C	X	X	X	X	Q_D
Q_C	0	1	0	1	$\overline{Q_D}$
$\overline{Q_C}$	0	1	0	1	$\overline{Q_D}$
$\overline{Q_C}$	X	1	0	X	Q_D
	Q_B	$\overline{Q_B}$		Q_B	

2 $D_B = Q_A\overline{Q_B} + \overline{Q_A}Q_B$

	Q_A		$\overline{Q_A}$		
Q_C	X	X	X	X	Q_D
Q_C	0	1	1	1	$\overline{Q_D}$
$\overline{Q_C}$	1	0	0	0	$\overline{Q_D}$
$\overline{Q_C}$	X	0	0	X	Q_D
	Q_B	$\overline{Q_B}$		Q_B	

3 $D_C = Q_A Q_B \overline{Q_C} + \overline{Q_B} Q_C + \overline{Q_A} Q_C$

	Q_A		$\overline{Q_A}$		
Q_C	X	X	X	X	Q_D
Q_C	1	0	0	0	$\overline{Q_D}$
$\overline{Q_C}$	0	0	0	0	$\overline{Q_D}$
$\overline{Q_C}$	X	1	1	X	Q_D
	Q_B	$\overline{Q_B}$		Q_B	

4 $D_D = Q_D + Q_A Q_B Q_C$

These will map Q_A, Q_B, Q_C and Q_D for each D value. The mapped equations are then simplified by looping the appropriate squares with the results shown.

Use of J-K Flip-Flops

The design of a decade counter using J-K flip-flops follows the same sequence as for the D flip-flops.

Using the J-K flip-flop transition table (Table 5.4) the required J and K conditions needed to cause each required transition can be filled in the third columns of Table 5.6. For example, if the present count is 2 = 0010, the next count is 3 = 0011, and the required J and K inputs are

Table 5.6 Decade counter state table (J-K flip-flops)

	Present state					*Next state*				*Required J and K inputs*							
	Q_D	Q_C	Q_B	Q_A		Q_D^+	Q_C^+	Q_B^+	Q_A^+	J_D	K_D	J_C	K_C	J_B	K_B	J_A	K_A
0	0	0	0	0	1	0	0	0	1	0	X	0	X	0	X	1	X
1	0	0	0	1	2	0	0	1	0	0	X	0	X	1	X	X	1
2	0	0	1	0	3	0	0	1	1	0	X	0	X	X	0	1	X
3	0	0	1	1	4	0	1	0	0	0	X	1	X	X	1	X	1
4	0	1	0	0	5	0	1	0	1	0	X	X	0	0	X	1	X
5	0	1	0	1	6	0	1	1	0	0	X	X	0	1	X	X	1
6	0	1	1	0	7	0	1	1	1	0	X	X	0	X	0	1	X
7	0	1	1	1	8	1	0	0	0	1	X	X	1	X	1	X	1
8	1	0	0	0	9	1	0	0	1	X	0	0	X	0	X	1	X
9	1	0	0	1	10	0	0	0	0	X	1	0	X	0	X	X	1
10	X	X	X	X		X	X	X	X	X	X	X	X	X	X	X	X
↓																	
15	X	X	X	X		X	X	X	X	X	X	X	X	X	X	X	X

a) Q_A must change from 0 to 1, hence $J_A = 1$, $K_A = X$.

b) Q_B must remain at 1, hence $J_B = X$, $K_B = 0$.

c) Q_C and Q_D must remain at 0, hence $J_C = J_D = 0$ and $K_C = K_D = X$.

Note that, when the next state is a don't-care, all the J and K inputs are also don't-cares.

The next step is to draw a number of state diagrams, one for each J input and for each K input. These diagrams map Q_A, Q_B, Q_C and Q_D for each J or K value. The mapped equations are simplified by looping the appropriate squares with the results shown.

The designed circuits are given in Figs. 5.20(*a*) and (*b*).

The designs can be checked using a *state checking table* (Table 5.7). Consider the J-K counter.

1 The count starts at 0 so write $Q_A = Q_B = Q_C = Q_D = 0$.

	$Q_A Q_B$	$Q_A \bar{Q}_B$	$\bar{Q}_A \bar{Q}_B$	$\bar{Q}_A Q_B$
$Q_C Q_D$	X	X	X	X
$Q_C \bar{Q}_D$	X	X	1	1
$\bar{Q}_C \bar{Q}_D$	X	X	1	1
$\bar{Q}_C Q_D$	X	X	1	X

1 $J_A = 1$

	$Q_A Q_B$	$Q_A \bar{Q}_B$	$\bar{Q}_A \bar{Q}_B$	$\bar{Q}_A Q_B$
$Q_C Q_D$	X	X	X	X
$Q_C \bar{Q}_D$	1	1	X	X
$\bar{Q}_C \bar{Q}_D$	1	1	X	X
$\bar{Q}_C Q_D$	X	1	X	X

2 $K_A = 1$

	$Q_A Q_B$	$Q_A \bar{Q}_B$	$\bar{Q}_A \bar{Q}_B$	$\bar{Q}_A Q_B$
$Q_C Q_D$	X	X	X	X
$Q_C \bar{Q}_D$	X	1	0	X
$\bar{Q}_C \bar{Q}_D$	X	1	0	X
$\bar{Q}_C Q_D$	X	0	0	X

3 $J_B = Q_A \bar{Q}_D$

	$Q_A Q_B$	$Q_A \bar{Q}_B$	$\bar{Q}_A \bar{Q}_B$	$\bar{Q}_A Q_B$
$Q_C Q_D$	X	X	X	X
$Q_C \bar{Q}_D$	1	X	X	0
$\bar{Q}_C \bar{Q}_D$	1	X	X	0
$\bar{Q}_C Q_D$	X	X	X	X

4 $K_B = Q_A$

	$Q_A Q_B$	$Q_A \bar{Q}_B$	$\bar{Q}_A \bar{Q}_B$	$\bar{Q}_A Q_B$
$Q_C Q_D$	X	X	X	X
$Q_C \bar{Q}_D$	X	X	X	X
$\bar{Q}_C \bar{Q}_D$	1	0	0	0
$\bar{Q}_C Q_D$	X	0	0	X

5 $J_C = Q_A Q_B$

	$Q_A Q_B$	$Q_A \bar{Q}_B$	$\bar{Q}_A \bar{Q}_B$	$\bar{Q}_A Q_B$
$Q_C Q_D$	X	X	X	X
$Q_C \bar{Q}_D$	1	0	0	0
$\bar{Q}_C \bar{Q}_D$	X	X	X	X
$\bar{Q}_C Q_D$	X	X	X	X

6 $K_C = Q_A Q_B$

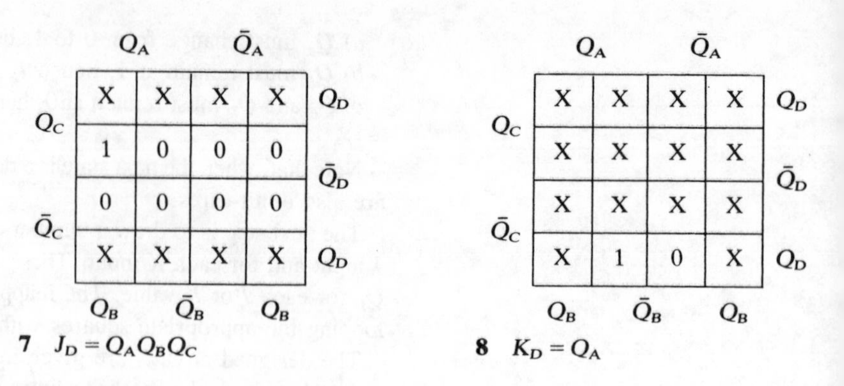

7 $J_D = Q_A Q_B Q_C$ 8 $K_D = Q_A$

Table 5.7 State checking table

	Present state												*Next state*				
	Q_D	Q_C	Q_B	Q_A	J_D	K_D	J_C	K_C	J_B	K_B	J_A	K_A	Q_D	Q_C	Q_B	Q_A	
0	0	0	0	0	0	0	0	0	0	0	1	1	0	0	0	1	1
1	0	0	0	1	0	1	0	0	1	1	1	1	0	0	1	0	2
2	0	0	1	0	0	0	0	0	0	0	1	1	0	0	1	1	3
3	0	0	1	1	0	1	1	1	1	1	1	1	0	1	0	0	4
4	0	1	0	0	0	0	0	0	0	0	1	1	0	1	0	1	5
5	0	1	0	1	0	1	0	0	1	1	1	1	0	1	1	0	6
6	0	1	1	0	0	0	0	0	0	0	1	1	0	1	1	1	7
7	0	1	1	1	1	1	1	1	1	1	1	1	1	0	0	0	8
8	1	0	0	0	0	0	0	0	0	0	1	1	1	0	0	1	9
9	1	0	0	1	0	1	0	0	0	1	1	1	0	0	0	0	0

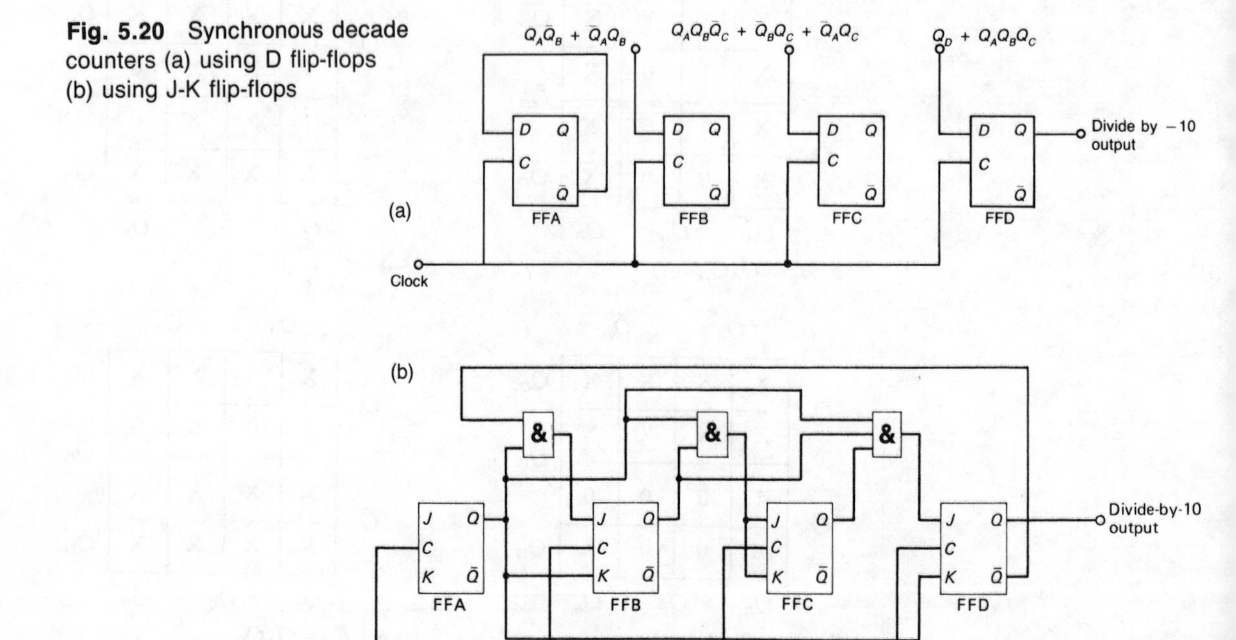

Fig. 5.20 Synchronous decade counters (a) using D flip-flops (b) using J-K flip-flops

2 Determine from the circuit the J and K values of each flip-flop at this time. Since $Q_A = 0$, then $J_B = K_B = J_C = K_C = J_D = K_D = 0$.

3 Now determine the next state. Since K_A and J_A are at 1, only the first stage will toggle so that the next state is 0001 or denary 1.

4 Write down the next state of the previous line as the present state of the next line and then find the values of each J and K input. Now $K_B = K_D = 1$ since $Q_A = 1$. Also, since $\bar{Q}_D = 1$, both inputs to the left-hand AND gate are 1 and thus $J_B = 1$. All other J and K inputs are at 0.

5 Determine the next state. Both flip-flops A and B toggle so that the next state is 0010 or denary 2.

6 This procedure is repeated, line-by-line, until a previous count is obtained. In this case this is 0000 which occurs on the tenth line.

Table 5.8

Present state					Next state					Required J-K inputs								Required D inputs			
	Q_D	Q_C	Q_B	Q_A		Q_D^+	Q_C^+	Q_B^+	Q_A^+	J_D	K_D	J_C	K_C	J_B	K_B	J_A	K_A	D_D	D_C	D_B	D_A
0	0	0	0	0	8	1	0	0	0	1	X	0	X	0	X	0	X	1	0	0	0
8	1	0	0	0	4	0	1	0	0	X	1	1	X	0	X	0	X	0	1	0	0
4	0	1	0	0	2	0	0	1	0	0	X	X	1	1	X	0	X	0	0	1	0
2	0	0	1	0	9	1	0	0	1	1	X	0	X	X	1	1	X	1	0	0	1
9	1	0	0	1	12	1	1	0	0	X	0	1	X	0	X	X	1	1	1	0	0
12	1	1	0	0	14	1	1	1	0	X	0	X	0	1	X	0	X	1	1	1	0
14	1	1	1	0	15	1	1	1	1	X	0	X	0	X	0	1	X	1	1	1	1
15	1	1	1	1	7	0	1	1	1	X	1	X	0	X	0	X	0	0	1	1	1
7	0	1	1	1	3	0	0	1	1	0	X	X	1	X	0	X	0	0	0	1	1
3	0	0	1	1	1	0	0	0	1	0	X	0	X	X	1	X	0	0	0	0	1
1	0	0	0	1	0	0	0	0	0	0	X	0	X	0	X	X	1	0	0	0	0

As a further example consider the design of a synchronous counter to follow the sequence 0, 8, 4, 2, 9, 12, 14, 15, 7, 3, 1, 0, etc. The state table for the required counter is given by Table 5.8.

For the other numbers, i.e. 5, 6, 10, 11, and 13 the D, or the J and K, inputs are all don't-cares.

From the state table, the state mappings for the D inputs are:

	Q_A		$\overline{Q_A}$		
Q_C	1	X	0	1	Q_D
Q_C	1	X	0	X	$\overline{Q_D}$
$\overline{Q_C}$	1	0	0	1	$\overline{Q_D}$
$\overline{Q_C}$	X	0	0	X	Q_D
	Q_B		$\overline{Q_B}$	Q_B	

1 $D_A = Q_B$

	Q_A		$\overline{Q_A}$		
Q_C	1	X	1	1	Q_D
Q_C	1	X	1	0	$\overline{Q_D}$
$\overline{Q_C}$	0	0	0	X	$\overline{Q_D}$
$\overline{Q_C}$	X	0	0	X	Q_D
	Q_B		$\overline{Q_B}$	Q_B	

2 $D_B = Q_AQ_C + \overline{Q_B}Q_C + Q_CQ_D$

	Q_A	Q_A	$\overline{Q}_A$	$\overline{Q}_A$	
Q_C	1	X	1	1	Q_D
Q_C	0	X	0	X	$\overline{Q}_D$
$\overline{Q}_C$	0	0	0	0	$\overline{Q}_D$
$\overline{Q}_C$	X	1	1	X	Q_D
	Q_B	$\overline{Q}_B$	$\overline{Q}_B$	Q_B	

3 $D_C = Q_D$

	Q_A	Q_A	$\overline{Q}_A$	$\overline{Q}_A$	
Q_C	0	X	1	1	Q_D
Q_C	0	X	0	1	$\overline{Q}_D$
$\overline{Q}_C$	0	0	1	X	$\overline{Q}_D$
$\overline{Q}_C$	X	1	0	X	Q_D
	Q_B	$\overline{Q}_B$	$\overline{Q}_B$	Q_B	

4 $D_D = Q_A\overline{Q}_CQ_D + \overline{Q}_AQ_B + \overline{Q}_AQ_CQ_D$

From the table the state mappings for each J and K input are as follows:

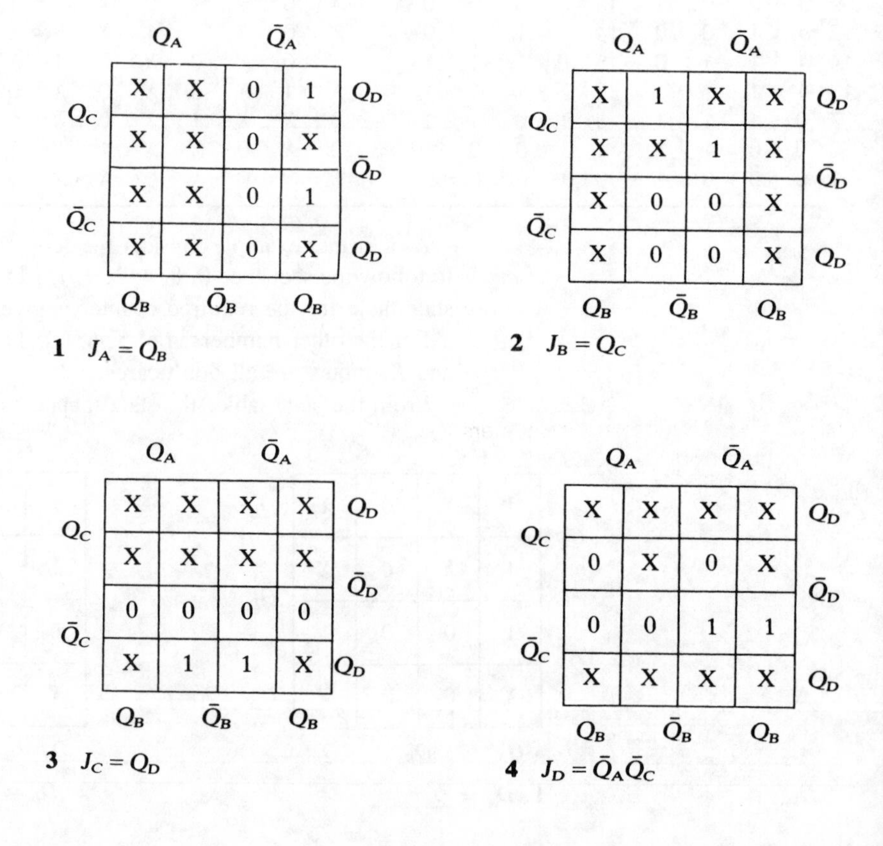

	Q_A	Q_A	$\bar{Q}_A$	$\bar{Q}_A$	
Q_C	X	X	0	1	Q_D
Q_C	X	X	0	X	$\bar{Q}_D$
$\bar{Q}_C$	X	X	0	1	$\bar{Q}_D$
$\bar{Q}_C$	X	X	0	X	Q_D
	Q_B	$\bar{Q}_B$	$\bar{Q}_B$	Q_B	

1 $J_A = Q_B$

	Q_A	Q_A	$\bar{Q}_A$	$\bar{Q}_A$	
Q_C	X	1	X	X	Q_D
Q_C	X	X	1	X	$\bar{Q}_D$
$\bar{Q}_C$	X	0	0	X	$\bar{Q}_D$
$\bar{Q}_C$	X	0	0	X	Q_D
	Q_B	$\bar{Q}_B$	$\bar{Q}_B$	Q_B	

2 $J_B = Q_C$

	Q_A	Q_A	$\bar{Q}_A$	$\bar{Q}_A$	
Q_C	X	X	X	X	Q_D
Q_C	X	X	X	X	$\bar{Q}_D$
$\bar{Q}_C$	0	0	0	0	$\bar{Q}_D$
$\bar{Q}_C$	X	1	1	X	Q_D
	Q_B	$\bar{Q}_B$	$\bar{Q}_B$	Q_B	

3 $J_C = Q_D$

	Q_A	Q_A	$\bar{Q}_A$	$\bar{Q}_A$	
Q_C	X	X	X	X	Q_D
Q_C	0	X	0	X	$\bar{Q}_D$
$\bar{Q}_C$	0	0	1	1	$\bar{Q}_D$
$\bar{Q}_C$	X	X	X	X	Q_D
	Q_B	$\bar{Q}_B$	$\bar{Q}_B$	Q_B	

4 $J_D = \bar{Q}_A\bar{Q}_C$

	$Q_A Q_B$	$Q_A \bar{Q}_B$	$\bar{Q}_A \bar{Q}_B$	$\bar{Q}_A Q_B$
$Q_C Q_D$	0	X	X	X
$Q_C \bar{Q}_D$	0	X	X	X
$\bar{Q}_C \bar{Q}_D$	0	1	X	X
$\bar{Q}_C Q_D$	X	1	X	X

5 $K_A = \bar{Q}_B$

	$Q_A Q_B$	$Q_A \bar{Q}_B$	$\bar{Q}_A \bar{Q}_B$	$\bar{Q}_A Q_B$
$Q_C Q_D$	0	X	X	0
$Q_C \bar{Q}_D$	0	X	X	X
$\bar{Q}_C \bar{Q}_D$	1	X	X	1
$\bar{Q}_C Q_D$	X	X	X	X

6 $K_B = \bar{Q}_C$

	$Q_A Q_B$	$Q_A \bar{Q}_B$	$\bar{Q}_A \bar{Q}_B$	$\bar{Q}_A Q_B$
$Q_C Q_D$	0	X	0	0
$Q_C \bar{Q}_D$	1	X	1	X
$\bar{Q}_C \bar{Q}_D$	X	X	X	X
$\bar{Q}_C Q_D$	X	0	X	X

7 $K_C = \bar{Q}_D$

	$Q_A Q_B$	$Q_A \bar{Q}_B$	$\bar{Q}_A \bar{Q}_B$	$\bar{Q}_A Q_B$
$Q_C Q_D$	1	X	0	0
$Q_C \bar{Q}_D$	X	X	X	X
$\bar{Q}_C \bar{Q}_D$	X	X	X	X
$\bar{Q}_C Q_D$	X	0	1	X

8 $K_D = Q_A Q_C + \bar{Q}_A \bar{Q}_C$

The required circuits are shown in Figs. 5.21(*a*) and (*b*).

The state table for a divide-by-5 resettable counter using J-K flip-flops is given by Table 5.9.

Mapping and simplifying as before gives

$$J_A = \bar{Q}_C \bar{R} \quad K_A = 1 \quad J_B = Q_A \bar{R} \quad K_B = Q_A + R$$

$$J_C = Q_A Q_B \bar{R} \quad K_C = 1$$

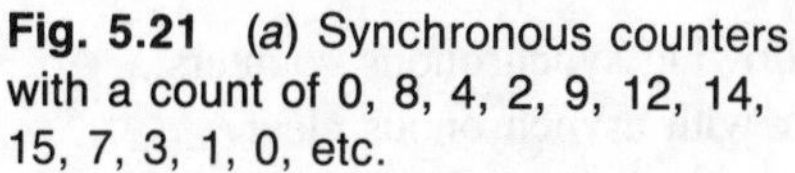

Fig. 5.21 (*a*) Synchronous counters with a count of 0, 8, 4, 2, 9, 12, 14, 15, 7, 3, 1, 0, etc.

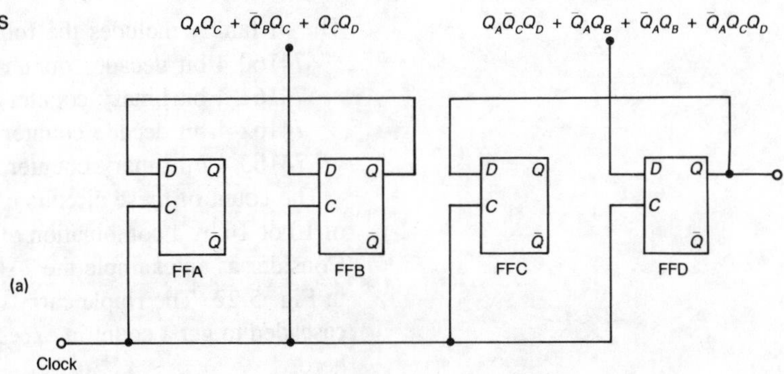

Fig. 5.21 (*b*)

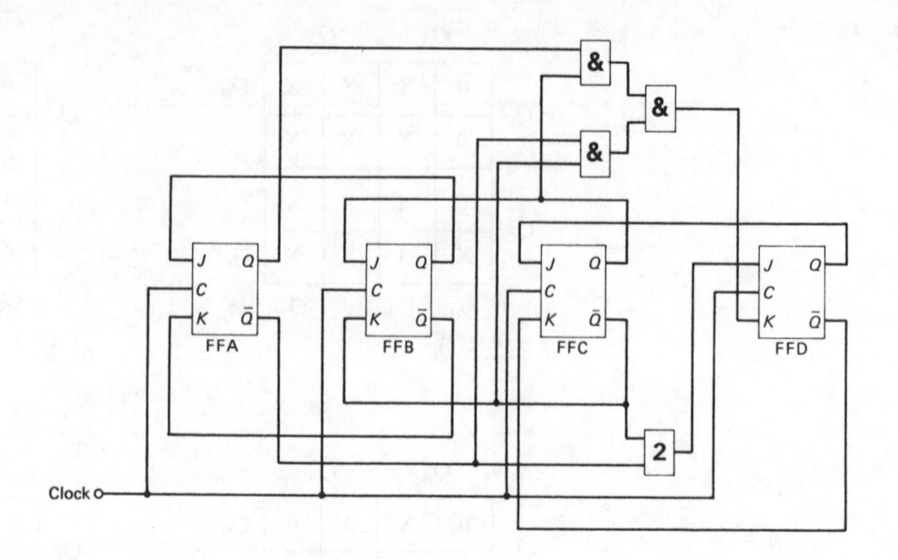

Table 5.9

Present state Q_C	Q_B	Q_A	Next state Q_C	Q_B	Q_A	Reset R	Required J-K inputs J_C	K_C	J_B	K_B	J_A	K_A
0	0	0	0	0	1	0	0	X	0	X	1	X
0	0	0	0	0	0	1	0	X	0	X	0	X
0	0	1	0	1	0	0	0	X	1	X	X	1
0	0	1	0	0	0	1	0	X	0	X	X	1
0	1	0	0	1	1	0	0	X	X	0	1	X
0	1	0	0	0	0	1	0	X	X	1	0	X
0	1	1	1	0	0	0	1	X	X	1	X	1
0	1	1	0	0	0	1	0	X	X	1	X	1
1	0	0	0	0	0	0	X	1	0	X	0	X
1	0	0	0	0	0	1	X	1	0	X	0	X

Synchronous Integrated Circuit Counters

The ttl family includes the following synchronous counters:

74160 4-bit decade counter with asynchronous clear.
74161 4-bit binary counter with asynchronous clear.
74162 4-bit decade counter with synchronous clear.
74163 4-bit binary counter with synchronous clear.

The count of these circuits can be reduced below the basic count of 10 or 16 by a combination of presetting and resetting or clearing. Consider as an example the 74161 whose pin connections are given in Fig. 5.22. The ripple carry output is used when the counters are cascaded to get a count in excess of 16 and it will not be considered here.

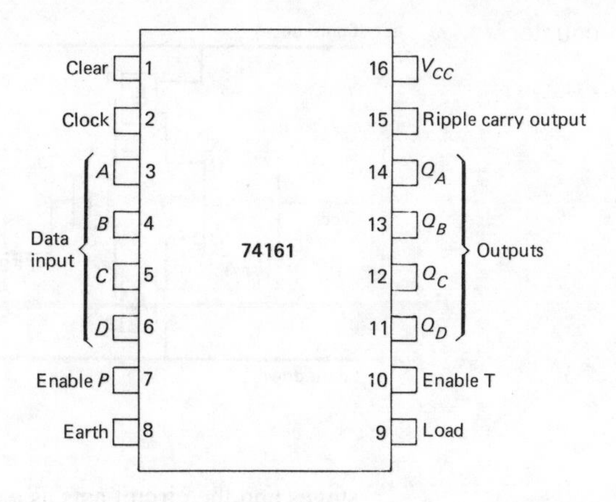

Fig. 5.22 Pin connections of 74161 4-bit binary counter

The counter is fully programmable so that all four of its outputs can be preset to be either 1 or 0. If the load terminal is held low, the counter is disabled and the outputs will agree with the input data after the next clock pulse. A low level at the clear input will clear or reset all the outputs to 0 regardless of the levels at the clock, enable P or T, or load terminals. Once the input data has been loaded, the circuit will start to count as soon as *both* the enable inputs are high. Once the wanted count has been reached, the counter can be inhibited by taking either or both enable inputs low.

Suppose that the counter is to be employed as a divide-by-8 circuit. The counter could be preset to binary 10, and it would then count through 11, 12, 13, 14, 15, 0, 1, 2 and then be disabled by the enable P input going low.

Synchronous counters in the cmos family are of the *Johnson* type and these are discussed later in this chapter.

Up–Down Counters

For some digital applications it is necessary to be able to count downwards, i.e. 9, 8, 7, 6, 5, 4, 3, 2, 1, and 0. Many circuits are able to count in either direction and are known as **up–down counters**.

It was mentioned on page 123 that a non-synchronous counter would count down if the $\bar{Q}$ output of each stage were connected to the clock input of the next stage. The change-over from one function to the other is arranged by means of AND gates that are enabled by a *count-up* or a *count-down* pulse.

The circuit of a **non-synchronous** 3-bit **up–down counter** is shown in Fig. 5.23. If the count-up line is taken to the logical 1 level, the AND gates *A* and *D* are enabled and connect the *Q* outputs of the flip-flops *A* and *B* to the clock input of the following flip-flops. The circuit then operates as an up-counter with a count of 8.

When the count-down line is taken to logical 1, gates *B* and *E* are enabled, while gates *A* and *D* are inhibited. Now the $\bar{Q}$ outputs of the flip-flops *A* and *B* are connected to the clock inputs of the following

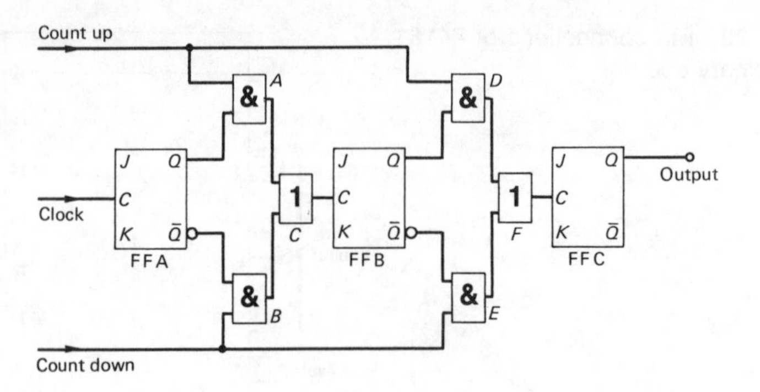

Fig. 5.23 Up-down counter

stages and the circuit acts as a down-counter, again with a count of 8.

The design of a **synchronous up–down counter** follows similar lines to that for an up-counter. Suppose, for example, that a divide-by-7 up–down counter is wanted. Assume that, when the count-up line is at 1, the count-down line will be at 0 and vice versa.

The state table of the required counter is given in Table 5.10. From the state table the state diagrams for each of the six J and K inputs can be written down (Z denotes the state of the count-up line):

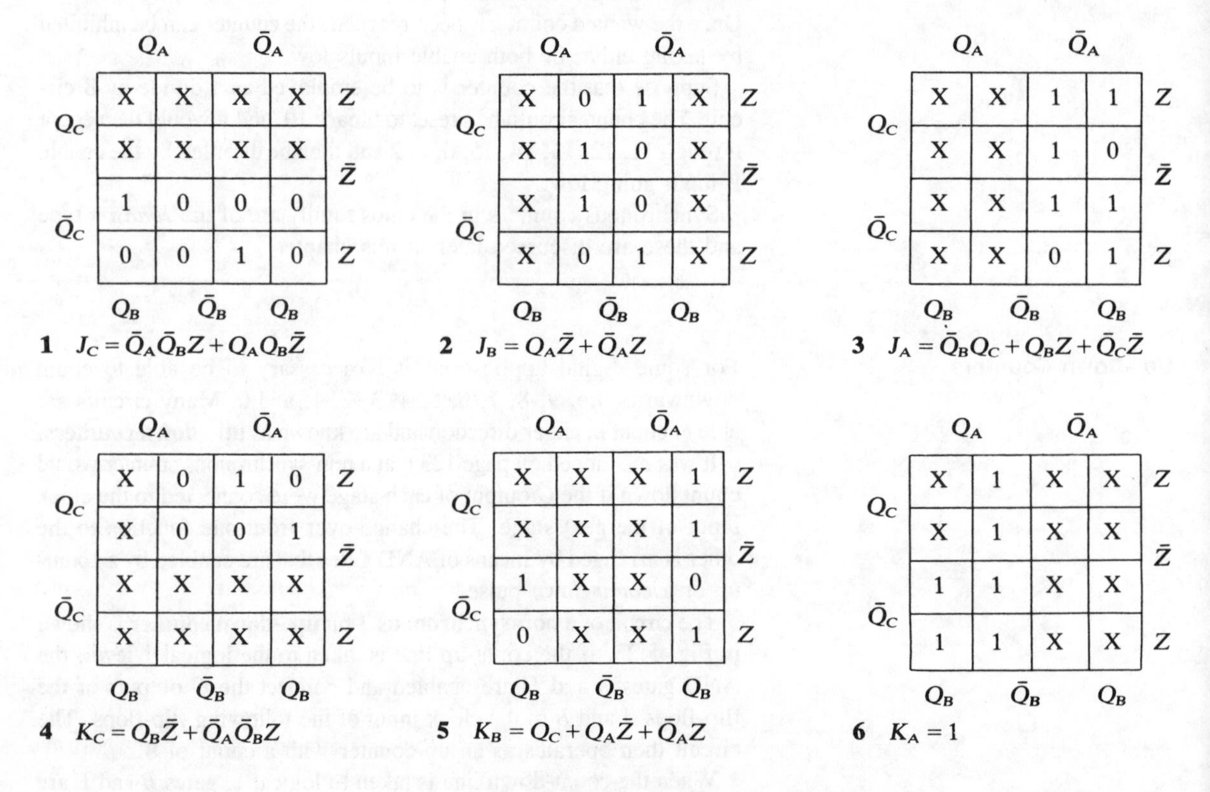

1 $J_C = \bar{Q}_A\bar{Q}_B Z + Q_A Q_B \bar{Z}$

2 $J_B = Q_A\bar{Z} + \bar{Q}_A Z$

3 $J_A = \bar{Q}_B Q_C + Q_B Z + \bar{Q}_C\bar{Z}$

4 $K_C = Q_B\bar{Z} + \bar{Q}_A\bar{Q}_B Z$

5 $K_B = Q_C + Q_A\bar{Z} + \bar{Q}_A Z$

6 $K_A = 1$

The required circuit is shown in Fig. 5.24.

Synchronous up–down counters available in the ttl family are the

Table 5.10 State table for divide-by-7 up–down counter

Present state			*Count-up line*	*Next state*			*Required J-K inputs*					
Q_C	Q_B	Q_A	Z	Q_C^+	Q_B^+	Q_A^+	J_C	K_C	J_B	K_B	J_A	K_A
0	0	0	0	0	0	1	0	X	0	X	1	X
0	0	0	1	1	1	0	1	X	1	X	0	X
0	0	1	0	0	1	0	0	X	1	X	X	1
0	0	1	1	0	0	0	0	X	0	X	X	1
0	1	0	0	0	1	1	0	X	X	0	1	X
0	1	0	1	0	0	1	0	X	X	1	1	X
0	1	1	0	1	0	0	1	X	X	1	X	1
0	1	1	1	0	1	0	0	X	X	0	X	1
1	0	0	0	1	0	1	X	0	0	X	1	X
1	0	0	1	0	1	1	X	1	1	X	1	X
1	0	1	0	1	1	0	X	0	1	X	X	1
1	0	1	1	1	0	0	X	0	0	X	X	1
1	1	0	0	0	0	0	X	1	X	1	0	X
1	1	0	1	1	0	1	X	0	X	1	1	X

Fig. 5.24

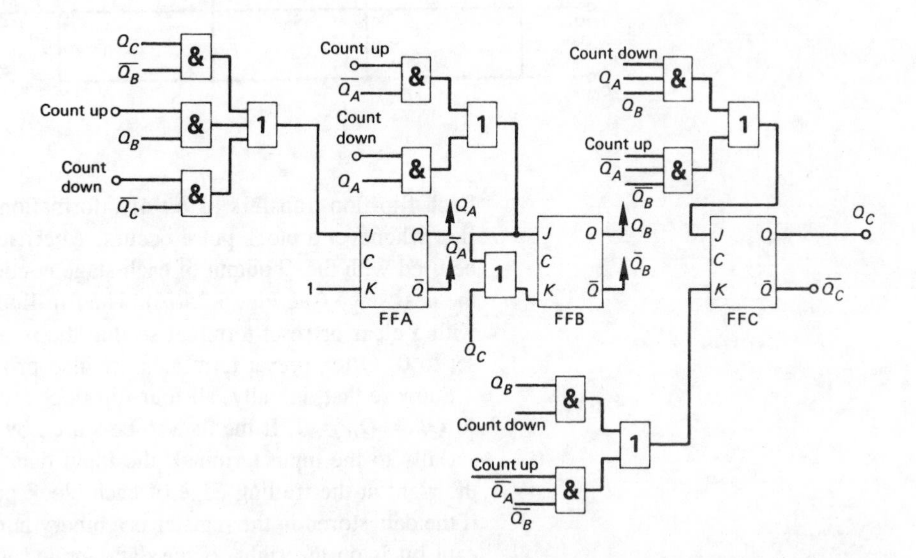

74190/1/2/3. These are, respectively, decade, binary, decade, and 4-bit binary counters and they are manufactured in both the standard and the low-power Schottky versions. The first two circuits incorporate ripple clocks but the other two have separate up–down clocks.

Shift Registers

A **shift register** consists of S-R or J-K flip-flops connected in cascade as shown by Fig. 5.25(*a*). The connection of Q_A to J_B and of $\bar{Q}_A$ to K_B and so on ensures that each flip-flop will take up the state, $Q = 0$ *or* $Q = 1$, of the preceding flip-flop at the end of each clock pulse.

Fig. 5.25 (*a*) Shift register using J-K flip-flops, (*b*) shift register using D flip-flops

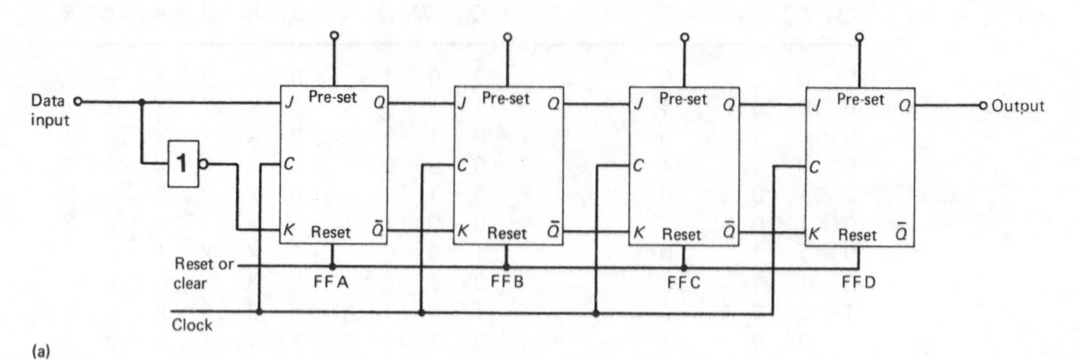

(b)

Each flip-flop transfers its *bit* of information to the following flip-flop whenever a clock pulse occurs. Alternatively, D flip-flops can be used with the Q output of each stage connected to the D input of the next stage (see Fig. 5.25(*b*)). The flip-flops are usually provided with a clear or reset terminal so that the register can be cleared or set to 0. Often preset terminals are also provided.

Suppose that, initially, all four flip-flops are cleared, i.e. $Q_A = Q_B = Q_C = Q_D = 0$. If the data to be stored by the register is applied serially to the input terminal, the input data will shift one stage to the right at the trailing edge of each clock pulse. This means that, if the data stored in the register is a binary number, the most significant bit is on the right. If the data applied to the input terminal is 1101, the bits 11 are loaded first and the action of the circuit is as follows.

At the end of the first clock pulse, $Q_A = 1$. After the second pulse, $Q_A = 1$, $Q_B = 1$. The third clock pulse will cause $Q_B = Q_C = 1$ but the first flip-flop will now reset to have $Q_A = 0$. The fourth bit of data is a 1 and this will be stored by flip-flop A at the end of the fourth clock pulse, when the bits stored by the other flip-flops all move one place to the right. At this stage the data stored is 1101. Now the complete number is stored by the register. At the end of the fifth clock pulse, flip-flop A clears and all the other flip-flops take up the state of the preceding stage. The most significant bit of the data has been shifted out of the register and lost.

After the sixth clock pulse, the two most significant bits have been lost, and so on. After eight pulses all the data has been lost.

The timing diagram illustrating the operation of the register is shown

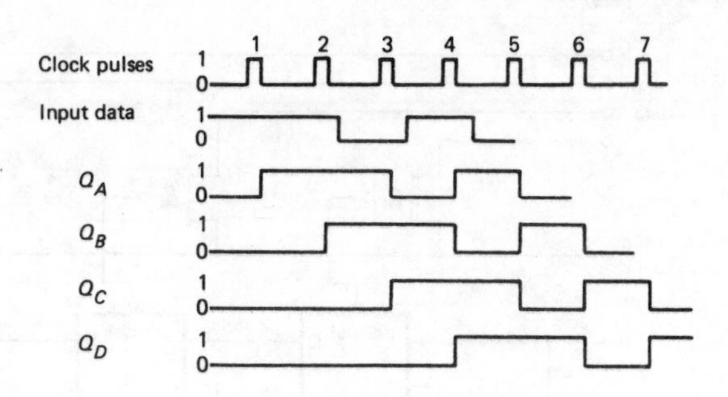

Fig. 5.26 Timing diagram for a 4-bit shift register

in Fig. 5.26. Note that the Q_A waveform is delayed behind the input data waveform by a time that may not be equal (in the figure it isn't) to the periodic time of the clock waveform. However, the Q waveforms of all the other flip-flops are delayed behind the Q waveform of the preceding flip-flop by *exactly* the clock period.

The shift register can be used to multiply or divide a binary number by a factor of 2 by merely shifting the number one place to the right or to the left, respectively. In either case any part of the number must not be shifted right out of the register.

In general, a shift register can be used in any one of four different ways. These methods are known as

1 serial-in/parallel-out (SIPO)
2 parallel-in/serial-out (PISO)
3 serial-in/serial-out (SISO)
4 parallel-in/parallel-out (PIPO)

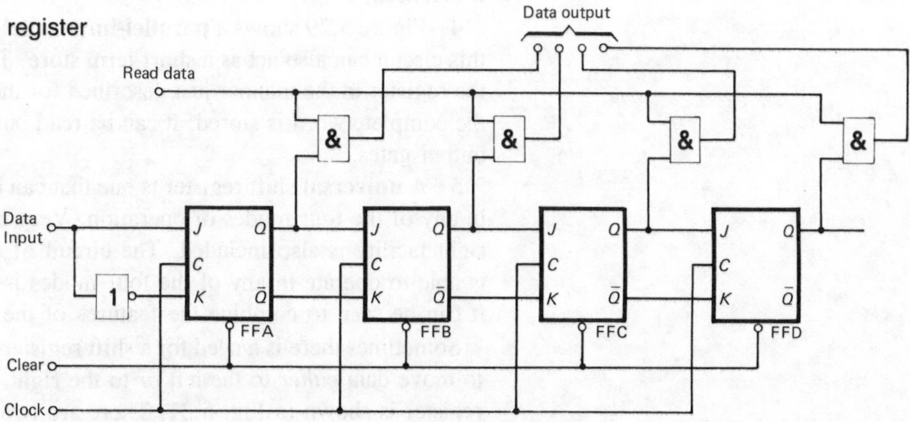

Fig. 5.27 SIPO shift register

1 With a **serial-in/parallel-out** shift register (Fig. 5.27), data is fed into the register in the manner previously described and, when the complete word is stored, all the bits can be read off simultaneously from the Q output of each stage by applying 1 to the read data line. The register acts to convert data from serial into parallel form.

Fig. 5.28 PISO shift register

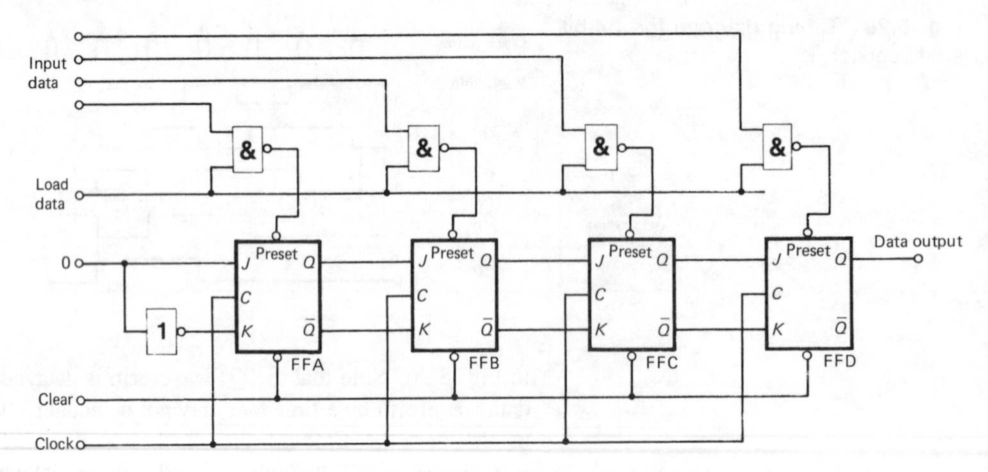

2 The **parallel-in/serial-out** shift register (Fig. 5.28) operates in exactly the opposite way. The data to be stored is set up by first clearing all the stages and then applying the data to the parallel input terminals.

When the data load line goes high, a 1 at a data input terminal will make the associated flip-flop's preset terminal go low. The flip-flops are assumed to have clear/reset terminals that are active low and so the flip-flop is set. If a data input is at 0, a high is applied to the flip-flop's preset terminal and the flip-flop is cleared.

3 The **serial-in/serial-out** shift register (Fig. 5.25) has already been described; it can be used as a delay circuit, or as a short-term store, but the stored data can only be accessed in the order in which it is stored.

4 Figure 5.29 shows a **parallel-in/parallel-out** shift register and this circuit can also act as a short-term store. The data is loaded into the register in the manner just described for the PISO circuit. When the complete word is stored, it can be read out by enabling the data output gates.

5 A **universal** shift register is one that can be programmed to act in any of the four modes of operation. Very often a shift-left/shift-right facility is also included. The circuit of a shift register which is able to operate in any of the four modes is shown by Fig. 5.30. It can be seen to combine the features of the previous circuits.

Sometimes there is a need for a shift register that has the capability to move data *either* to the left *or* to the right. The circuit of such a register is shown in Fig. 5.31. There are two data input terminals,

Fig. 5.29 PIPO shift register

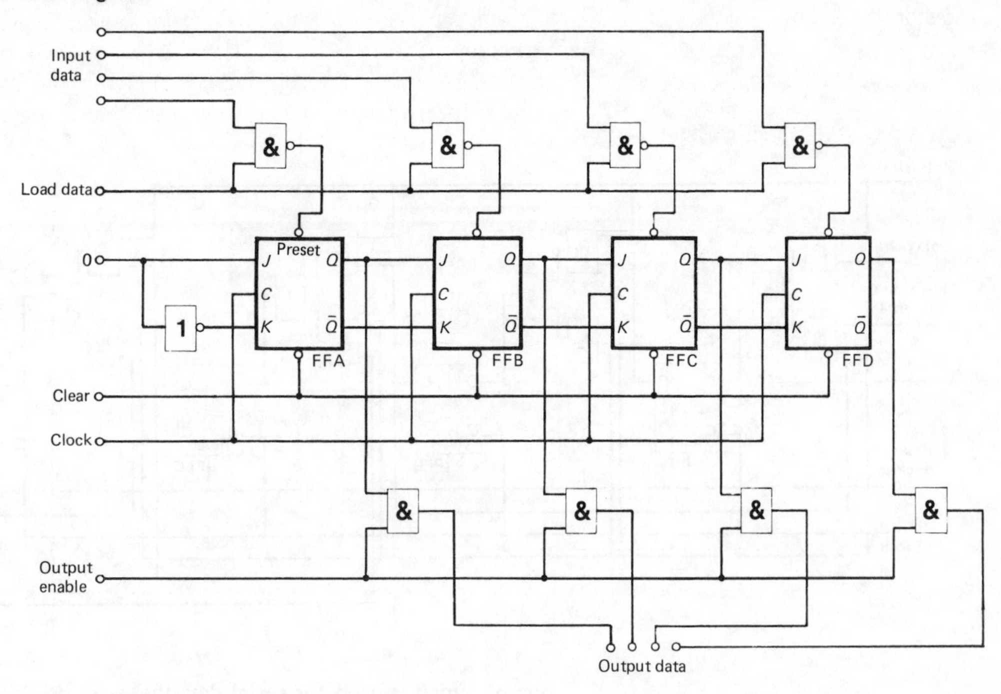

Fig. 5.30 Universal shift register

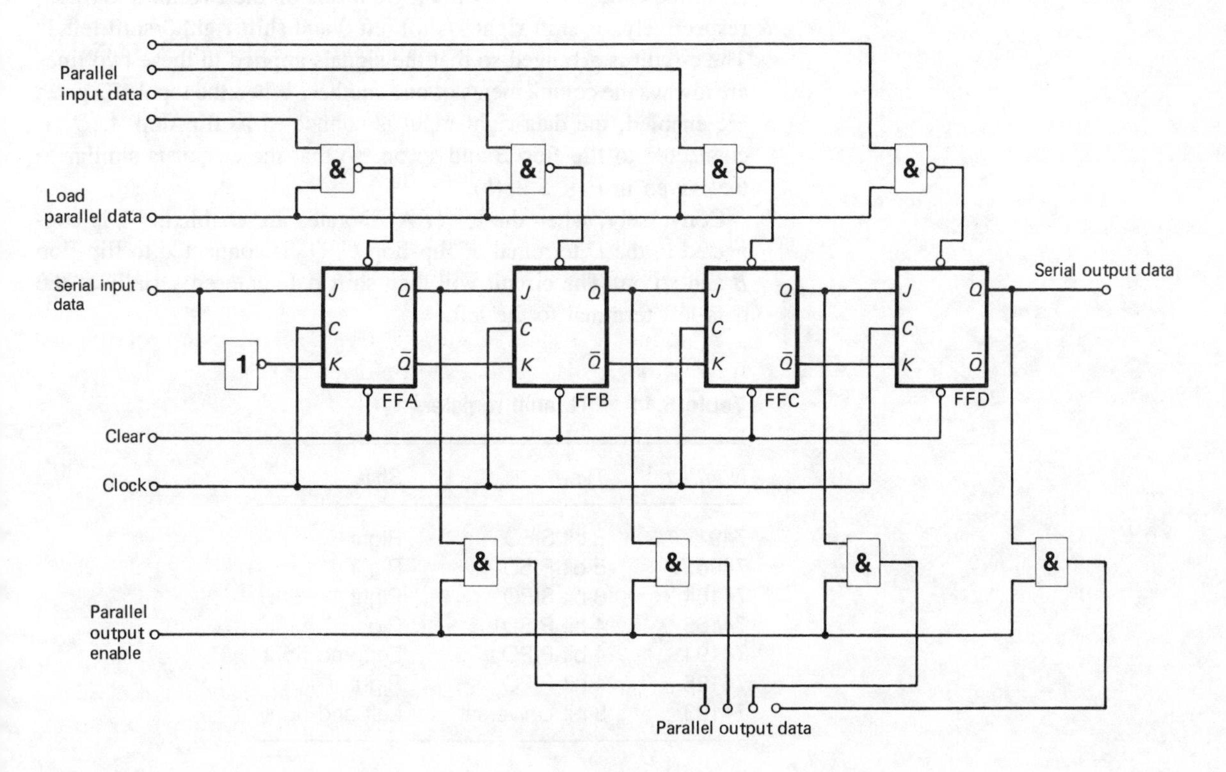

Fig. 5.31 Shift-right/shift-left register

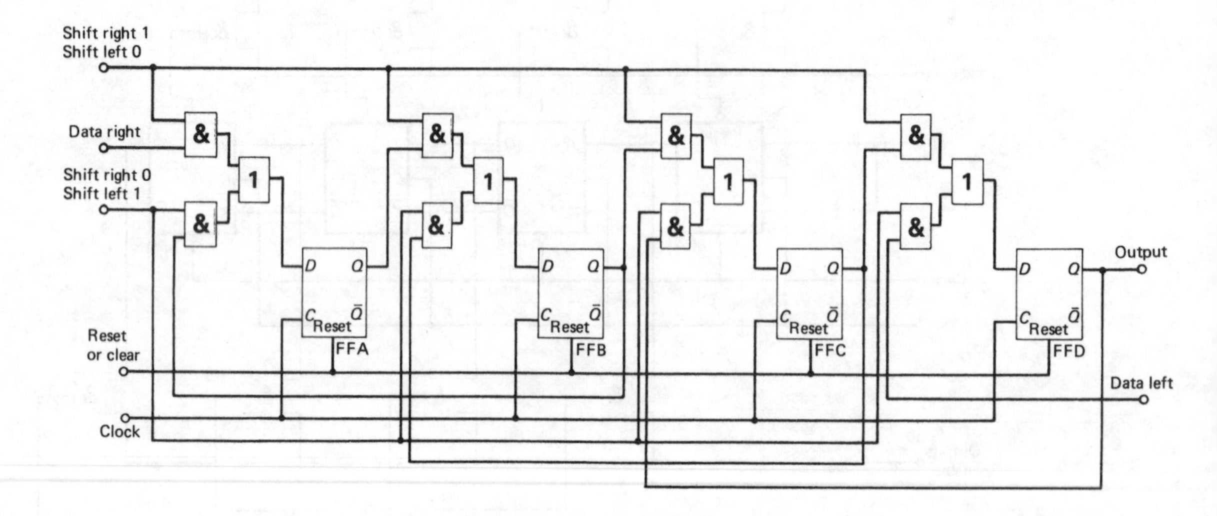

one of which is used for serial data that is to be shifted to the right and the other is for left-shifting data. The direction in which the data is shifted is determined by the logic levels on the two lines marked, respectively, as shift right 1/shift left 0 and shift right 0/shift left 1. The circuit is arranged so that the signals applied to these two lines are always the complements of one another. When the top AND gates are enabled, the data-right input is connected to flip-flop *A*. Q_A is connected to flip-flop *B* and so on, so that the circuit is similar to that given in Fig. 5.25(*b*).

Conversely, when the lower AND gates are enabled, Q_D is connected to the *D* terminal of flip-flop *C*, Q_C is connected to flip-flop *B* and so on. The circuit will then shift data entered serially at the data-left terminal to the left.

Table 5.11 TTL shift registers

Number	*Type*	*Shift*
7491	8-bit SISO	Right
7496	5-bit PISO	Right
74164	8-bit SIPO	Right
74165	4-bit PISO	Right
74194	4-bit PIPO	Left and right
74195	4-bit PIPO	Right
74198	8-bit Universal	Left and right

Shift registers are manufactured in both the ttl and the cmos logic families. Some examples in the ttl range are given in Table 5.11.

Cmos shift registers may be either static or dynamic types and examples of devices are given in Table 5.12.

Table 5.12 CMOS shift registers

Number	*Type*	*Shift*
4006	18-stage static SISO	Right
4014	8-stage static PISO or SISO	Right
4015	dual 4-bit static SIPO	Right
4021	8-stage static PISO or SISO	Right
4031	64-stage static SISO	Right
4035	4-stage static PIPO	Right
4062	200-stage dynamic SISO	Right
40100	32-stage static SISO	Left and right
40194	4-stage static Universal	Left and right

MOS Registers

Static shift registers store their data in a number of flip-flops and are able to hold the information for an indefinite length of time. A dynamic shift register, on the other hand, stores bits of information on the capacitances linking the stages of the register. Because the charge held in these capacitances gradually leaks away, the dynamic shift register must be periodically *refreshed*. The dynamic type of shift register therefore possesses the disadvantages of needing more complex supporting circuitry and a minimum clock frequency, but it does have the important advantage of a very much higher storage capability. This is made evident by Table 5.12 from which it can be seen that the 4062 dynamic shift register has 200 stages, compared with up to 64 stages for the static type. Other dynamic shift registers may have even larger storage capacities, for example, 512 stages or 1024 stages. Clearly, all large capacity shift registers must be of the serial-out type because the number of ic package pins is limited.

Dynamic cmos Shift Registers

Shift registers constructed by the interconnection of flip-flops as previously described are not practical when high bit storage capabilities are required. This is because the area of chip occupied would be too large and the power dissipation would be excessively high. Large capacity shift registers — perhaps hundreds of bits — are therefore always of the dynamic type.

Essentially, a dynamic shift register stage is fabricated by the cascade of two dynamic mos inverters, bit storage being *temporarily* achieved by charging the gate-substrate capacitance of a mosfet.

Fig. 5.32 Principle of a dynamic shift register

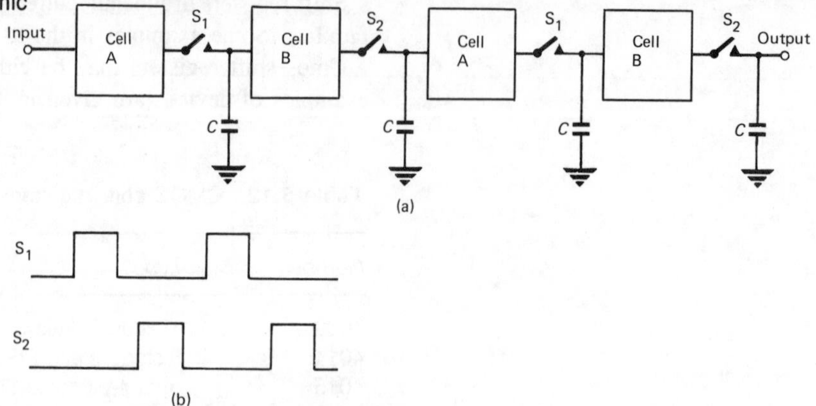

The principle of operation of a dynamic shift register is illustrated by Fig. 5.32. A number of mosfet memory cells are connected in cascade via switches S_1 and S_2 which are provided by transmission gates. Since a mosfet has a very high input impedance, the small stray input capacitance of a mosfet will be able to hold a charge for a relatively long period of time. Binary 1 is represented by the storage of charge and binary 0 is represented by zero charge being stored. Unfortunately, any stored charge will leak away unless it is continually *refreshed*. This is achieved by the circuit continually recirculating the stored data at a rate determined by the clock frequency.

Two clocks are used which close the switches S_1 and S_2 at different instants in time. When the switches S_1 are closed, the cells marked as A transfer the charge stored in their input capacitances to the input capacitances of the cells marked as B. Similarly, when the switches S_2 are closed (S_1 are then open), the data stored in the B cells is transferred to the A cells.

The maximum clock frequency is limited by the input time constant of the memory cells (because the capacitance must have sufficient time to charge up) with the associated switch closed. The minimum clock frequency is determined by the discharge time constant of a cell's input circuit.

Example 5.2

Figure 5.33 shows a part of the circuit of a dynamic shift register. The ON resistances of transistors T_1, T_2 and T_3 are, respectively, 3000 Ω, 100 kΩ, and 10 kΩ. If the stray capacitance C is equal to 4 pF, calculate *a)* the rise and fall times of the output (capacitor) voltage, and *b)* the maximum clock frequency.

Solution

a) The rise time is given by the expression

$$t_r = 2.5(R_2+R_3)\,C$$

Fig. 5.33

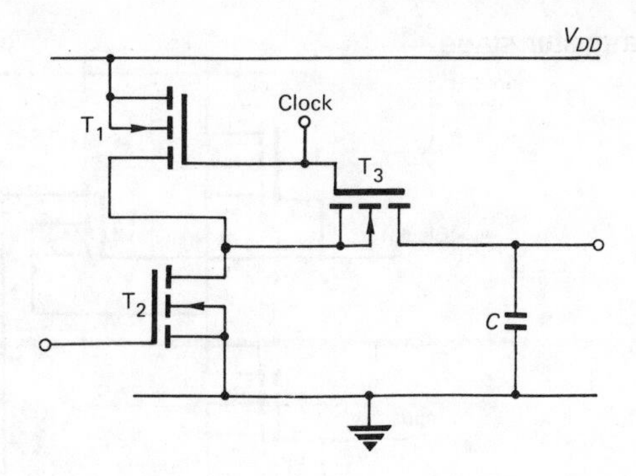

$$= 2.5\times110\times10^3\times4\times10^{-12} = 1.1\ \mu s \quad (Ans)$$

The fall time is given by

$$t_f = 2.5(R_1+R_3)C$$

$$= 2.5\times13\times10^3\times4\times10^{-12} = 0.13\ \mu s \quad (Ans)$$

b) The maximum clock frequency is

$$f_{max} = 1/2(2t_r+t_f) = 1/2\times2.33\times10^{-6} = 215\ \text{kHz} \quad (Ans)$$

The circuit of a **dynamic cmos shift register** is shown in Fig. 5.34. The boxes marked as TG represent *transmission gates* (p. 59) that are controlled by complementary clock signals. Transistors T_1 and T_2, and transistors T_3 and T_4, form two cmos inverter stages.

When the clock is at the logical 1 voltage level, TG_1 is turned ON and the input signal is applied to T_1/T_2 *and* across the capacitance C_1. When the clock becomes 0, TG_1 turns OFF and TG_2 turns ON; capacitances C_2 and C_3 are now effectively placed in parallel and the voltage developed across them is inverted by T_3/T_4 to appear across the output capacitance C_4. This means that the input signal appears at the output of the circuit after a time delay equal to the periodic time of the clock.

Fig. 5.34 CMOS dynamic shift register stage

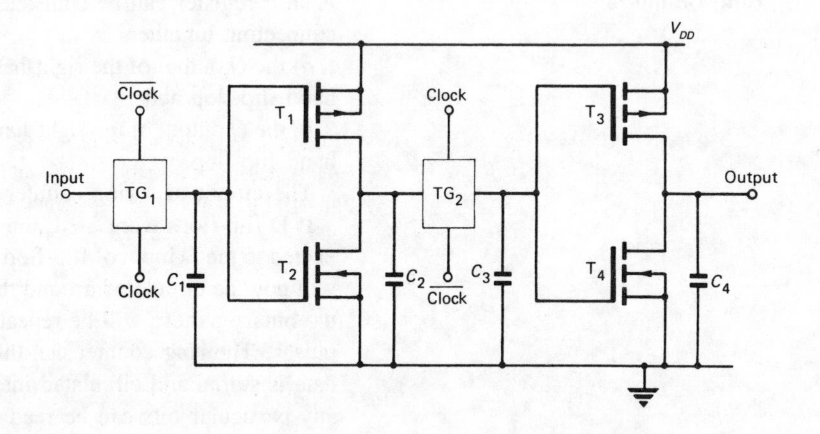

Fig. 5.35 NMOS shift register stage

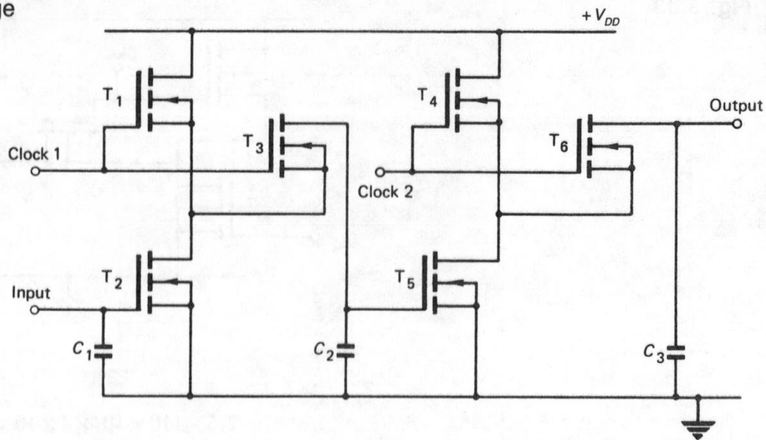

Although the cmos shift register contains eight mosfets in all, the total power dissipation is very small since the only currents that flow are the minute ones needed to charge the four capacitances. These capacitances are only of the order of 0.5 pF.

NMOS dynamic shift registers are also used when very large storage capacities are needed and these may well be able to store more than 1000 bits. Figure 5.35 gives the circuit of an nmos shift register stage; T_1 and T_4 act as the active loads. When a clock is at 1, the associated mosfets are turned ON and the voltage present at the gate of T_2, or T_5, will be inverted. The clocks are phased so that T_3 and T_6 will not be ON together. Since either T_3 or T_6 will be ON, the inverted voltage will charge up either capacitance C_2 or C_3. When clock 1 = 1, clock 2 = 0. Hence a new input bit stored in capacitance C_1 will not affect the output of the stage until clock 2 becomes 1.

Shift registers are commonly employed in conjunction with full-adders to produce arithmetic circuits (see p. 78).

Ring Counters

A shift register can be connected to function as a **ring counter** by connecting together

a) the Q output of the right-hand flip-flop to the J input of the left-hand flip-flop and

b) the $\bar{Q}$ output of the right-hand flip-flop to the K input of the left-hand flip-flop.

The circuit of a ring counter is shown in Fig. 5.36.

If D flip-flops were used, the Q output of flip-flop D will be connected to the D input of flip-flop A. The internal states of the register will now be circulated around the loop and, for an *n-stage register*, the binary pattern will be repeated at the output after every n clock pulses. The ring counter can therefore be used as a *delay line*; the data is stored and circulated internally in a multiple bit pattern and any particular bits can be read out as and when required.

Fig. 5.36 Ring counter

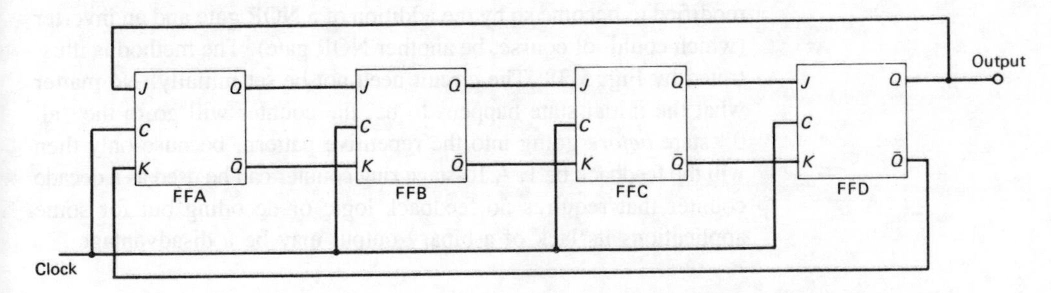

If only a single bit is circulated, the output of each flip-flop will give a uniquely timed 1-*out-of-n count*. The count is read by finding the flip-flop that is set; no decoding is needed. Consider, for example, a 5-stage register; the pattern held is given in Table 5.13 and the timing diagram is shown by Fig. 5.37(*a*). If the flip-flops *A and B* are initially set (Fig. 5.37(*b*)), a double-width pulse will circulate around the register.

Table 5.13 5-stage ring counter state table

Clock pulse	Q_E	Q_D	Q_C	Q_B	Q_A
	0	0	0	0	1
1	0	0	0	1	0
2	0	0	1	0	0
3	0	1	0	0	0
4	1	0	0	0	0
5	0	0	0	0	1

Fig. 5.37 Timing diagram for a 5-stage ring counter

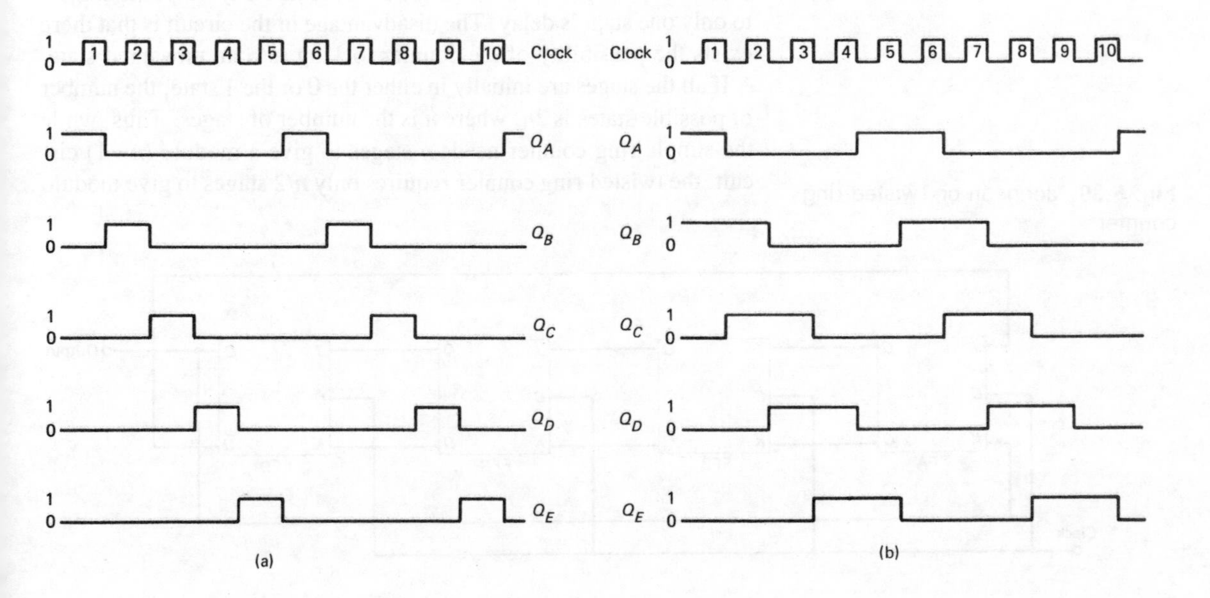

The basic ring counter is not self-starting but it can easily be modified to become so by the addition of a NOR gate and an inverter (which could, of course, be another NOR gate). The method is illustrated by Fig. 5.38. The circuit need not be set initially. No matter what the initial state happens to be, the counter will go to the "all 0" state *before* going into the repetitive pattern, because only then will the feedback be 1. A 10-stage ring counter can be used as a decade counter that requires no feedback logic or decoding but for some applications its lack of a binary output may be a disadvantage.

Fig. 5.38 Self-starting ring counter

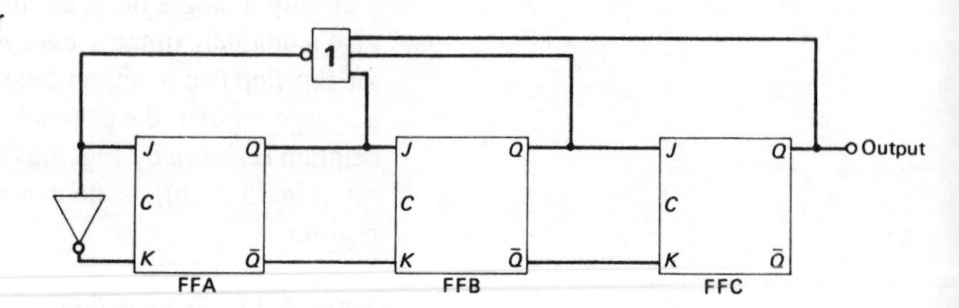

Johnson Counter

A **Johnson counter** or **twisted ring counter** differs from the ring counter in that the Q and $\bar{Q}$ connections of the final flip-flop are interchanged, or *twisted* (see Fig. 5.39). The sequence of the Johnson counter is given by Table 5.14. Decoding of each count is easy since it only needs one 2-input AND gate for each count.

The Johnson counter possesses the advantages of (i) high speed of operation because it does not follow the true 8421 binary code and (ii) the decoded outputs are glitch-free because any output is subject to only one stage's delay. The disadvantage of the circuit is that there exists the possibility of the counter locking into an unwanted state.

If all the stages are initially in either the 0 or the 1 state, the number of possible states is $2n$, where n is the number of stages. Thus, while the simple ring counter needs n stages to give a modulo $(n+1)$ circuit, the twisted ring counter requires only $n/2$ stages to give modulo

Fig. 5.39 Johnson or Twisted ring counter

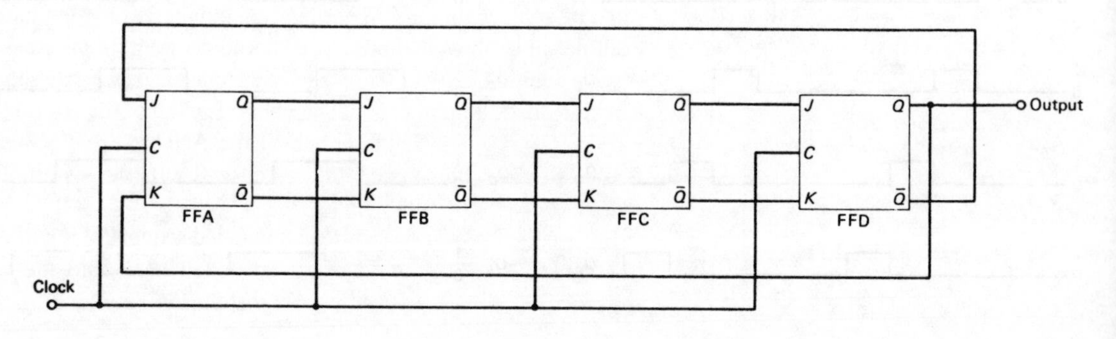

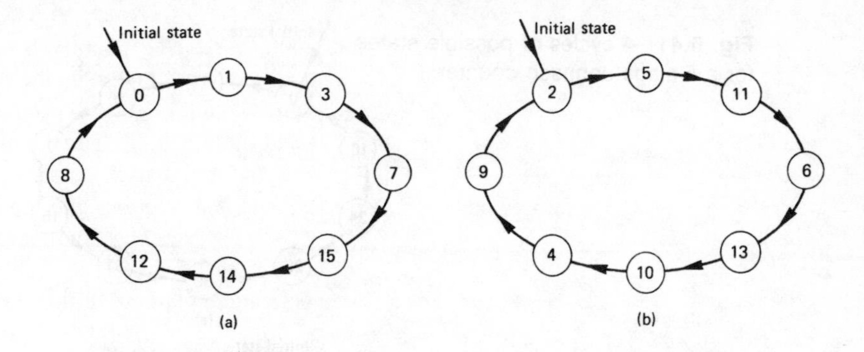

Fig. 5.40 Possible states for a 4-stage Johnson counter

n. Referring again to the 4-stage Johnson counter, the 8 possible states are shown in Fig. 5.40(*a*) when the initial count is 0 (as in Table 5.14) and in Fig. 5.40(*b*) when the initial count is 2.

Table 5.14 Johnson counter sequence

Count	Q_D	Q_C	Q_B	Q_A	*Inputs to the AND gate for decoded count*
0	0	0	0	0	$\bar{Q}_A \bar{Q}_D$
1	0	0	0	1	$Q_A \bar{Q}_B$
2	0	0	1	1	$Q_B \bar{Q}_C$
3	0	1	1	1	$Q_C \bar{Q}_D$
4	1	1	1	1	$Q_A Q_D$
5	1	1	1	0	$\bar{Q}_A Q_B$
6	1	1	0	0	$\bar{Q}_B Q_C$
7	1	0	0	0	$\bar{Q}_C Q_D$
8	0	0	0	0	$\bar{Q}_A \bar{Q}_D$

Similarly, the four groups of possible states for a 5-stage Johnson counter are given in Fig. 5.41. Notice that the *total* number of different states is 32 or 2^5. For the first cycle (Fig. 5.41(*a*)), the required output decoding is given in Table 5.15.

The Johnson counter is not self-starting unless some extra feedback circuitry is provided. Suitable feedback for this purpose can be designed.

To make the counter self-starting, logic representing one state from each possible unwanted sequence should be used. In the case of the 4-stage Johnson counter there is only the one other possible sequence. Thus, if, for example, decimal number 5 is chosen, the required logic is $F = Q_A\bar{Q}_BQ_C\bar{Q}_D$. If the output of the AND gate with this input is used to clear the counter, the circuit will self-start within 8 clock pulses.

If the counter should start in an unwanted sequence, it will remain in it unless logic is employed to switch it back into the wanted sequence. Consider again the 4-stage Johnson counter.

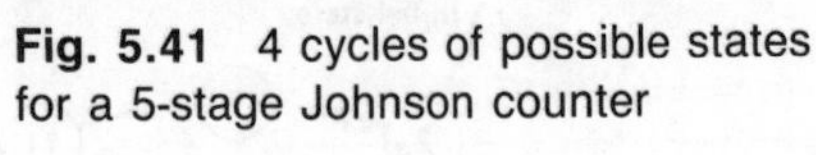

Fig. 5.41 4 cycles of possible states for a 5-stage Johnson counter

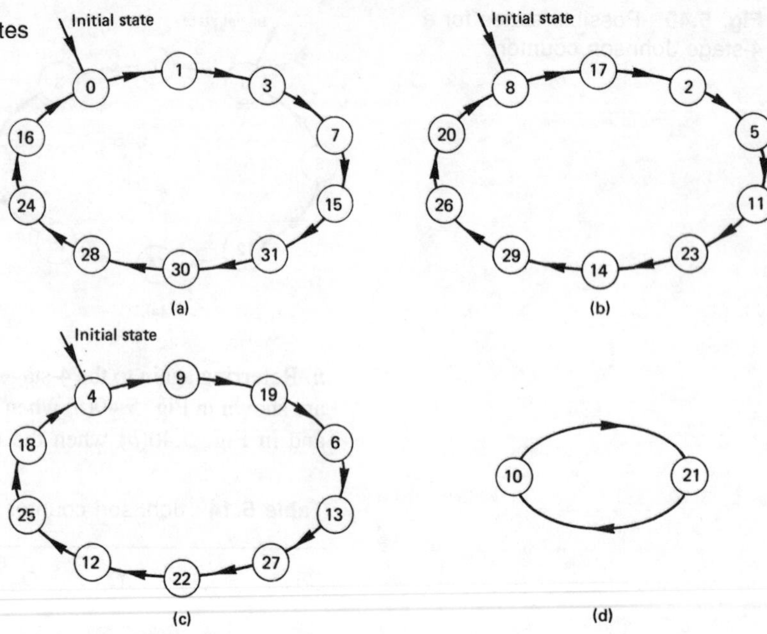

Table 5.15

Decimal number	*Count*	Q_E	Q_D	Q_C	Q_B	Q_A	*Inputs to AND gate for decoded count*
0	0	0	0	0	0	0	$\bar{Q}_A\,\bar{Q}_E$
1	1	0	0	0	0	1	$Q_A\,\bar{Q}_B$
2	3	0	0	0	1	1	$Q_B\,\bar{Q}_C$
3	7	0	0	1	1	1	$Q_C\,\bar{Q}_D$
4	15	0	1	1	1	1	$Q_D\,\bar{Q}_E$
5	31	1	1	1	1	1	$Q_A\,Q_E$
6	30	1	1	1	1	0	$\bar{Q}_A\,Q_B$
7	28	1	1	1	0	0	$\bar{Q}_B\,Q_C$
8	24	1	1	0	0	0	$\bar{Q}_C\,Q_D$
9	16	1	0	0	0	0	$\bar{Q}_D\,Q_E$

The Karnaugh map showing the decimal equivalents of the states in the wanted sequence for a 4-stage counter is

		Q_A	Q_A	$\bar{Q}_A$	$\bar{Q}_A$	
Q_C		15	U	12	14	Q_D
Q_C		7	U	U	U	$\bar{Q}_D$
$\bar{Q}_C$		3	1	0	U	$\bar{Q}_D$
$\bar{Q}_C$		U	U	8	U	Q_D
		Q_B	$\bar{Q}_B$	$\bar{Q}_B$	Q_B	

The squares representing unused states have been marked U (refer to Fig. 5.40). The Boolean expression that represents these states is

$$F = Q_A\bar{Q}_BQ_C + \bar{Q}_AQ_C\bar{Q}_D + Q_A\bar{Q}_CQ_D + \bar{Q}_AQ_B\bar{Q}_C$$

The logic of these states can be used to clear the counter if the wrong sequence is entered.

The cmos 4017 is a 5-stage Johnson decade counter that has ten decoded outputs (Fig. 5.42(*a*)), each of which can drive other circuits. If one of the outputs is connected to the clock inhibit terminal, the count can be stopped at any desired point. The outputs go high in sequence, one at a time, under the control of the clock. The terminal marked as Carry out goes low after five, and then after ten, clock pulses and can be connected to the clock terminal of another 4017 if a count greater than 10 is wanted. The reset terminal can be used to set the 0 output to 1 and hence all the other outputs to 0. The timing diagram of a 4017 is shown in Fig. 5.42(*b*).

Another useful presettable cmos Johnson counter is the 4018 (Fig. 5.42(*c*)). This is a 5-stage circuit that can be used to divide by any ratio from 2 to 10 inclusive. To divide by any *even* number, the $\bar{Q}_n$

Fig. 5.42 (*a*) Pin connections of the 4017 5-stage Johnson counter, (*b*) 4017 timing diagram, (*c*) Pin connections of the 4018 presettable 5-stage Johnson counter

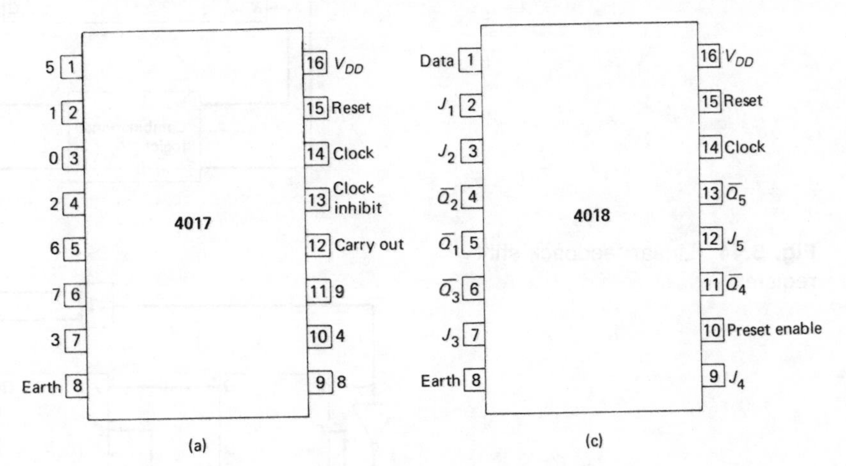

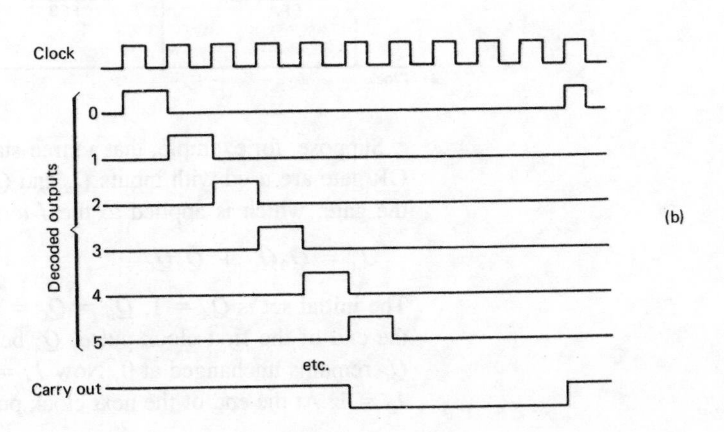

output, where n is *one-half* of the required ratio, is connected to the data terminal. To divide by any *odd* ratio, the $\bar{Q}_m$ and $\bar{Q}_n$ terminals, where $m+n$ = required ratio, are connected to the inputs of a 2-input AND gate and the output of this gate to the data terminal. The preset and enable signal allows data on the J inputs to reset the counter.

Feedback Shift Registers

In the ring and twisted ring counters the Q and the $\bar{Q}$ outputs of the final stage are taken directly back to the first stage. In a **feedback shift register** the feedback is passed through a combinational logic circuit (see Fig. 5.43). If the combinational logic circuit consists *only* of exclusive-OR gates, the circuit is said to be a *linear feedback shift register*.

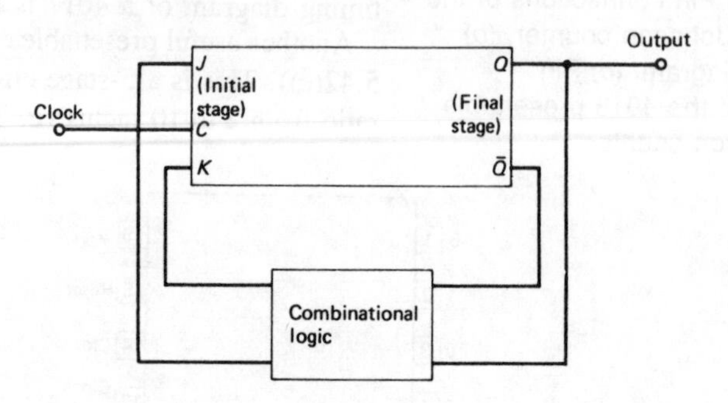

Fig. 5.43 Feedback shift register

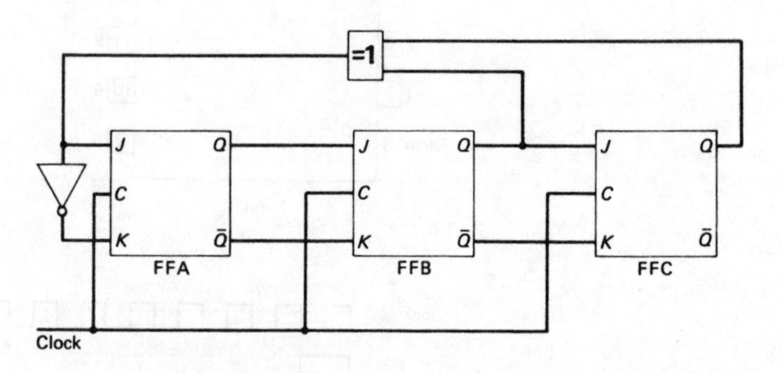

Fig. 5.44 Linear feedback shift register

Suppose, for example, that a three-stage circuit and one exclusive-OR gate are used with inputs Q_B and Q_C (Fig. 5.44). The output of the gate, which is applied to the J terminal of the first stage, is

$$J_A = Q_B\bar{Q}_C + \bar{Q}_B Q_C$$

The initial set is $Q_A = 1$, $Q_B = Q_C = 0$, hence $J_A = 0$, $J_B = 1$. At the end of the first clock pulse, Q_A becomes 0 and Q_B becomes 1, Q_C remains unchanged at 0. Now $J_A = 1.1+0.0 = 1$, $J_B = 0$, and $J_C = 1$. At the end of the next clock pulse, Q_A becomes 1 again, Q_B

= 0, and Q_C = 1. Now $J_A = 0.0+1.1 = 1$, $J_B = 1$, and $J_C = 0$, and so the next state is $Q_A = Q_B = 1$ and $Q_C = 0$. The operation is summarized by Table 5.16.

The "all Qs zero" state must be avoided at switch-on because it would lock the J_A input at 0. This requirement can be achieved by ANDing all the $\bar{Q}$ outputs and applying the output of the AND gate to the J_A input. This means that the maximum cycle length is $2^n - 1$. The stages from which the feedback must be taken to give a maximum cycle length are shown in Table 5.17.

Table 5.16 State table of linear feedback register

Q_C	Q_B	Q_A
0	0	1
0	1	0
1	0	1
0	1	1
1	1	1
1	1	0
1	0	0
0	0	1

Table 5.17

Number of stages	*Linear feedback*	*Cycle length* ($2^n - 1$)
2	1⊕2	3
3	2⊕3	7
4	3⊕4	15
5	3⊕5	31
6	5⊕6	63
7	6⊕7	127
8	4⊕3⊕4⊕8	255
9	5⊕9	511
10	7⊕10	1023

(Note: ⊕ indicates exclusive-OR function.)

Other feedback combinations will produce shorter cycle lengths.

The output of a linear fsr forms a pseudo-random binary sequence which, if there are a large enough number of stages, may approximate to the white noise power density spectrum. For example, from Table 5.16 it can be seen that the output of a 3-stage circuit is

1001101110111010100 01

This sequence contains all the possible 3-bit combinations *except* 000.

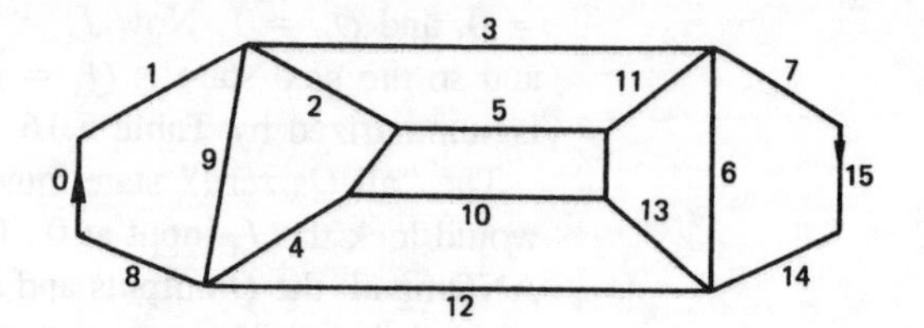

Fig. 5.45 Determination of the possible states in a non-linear feedback shift register

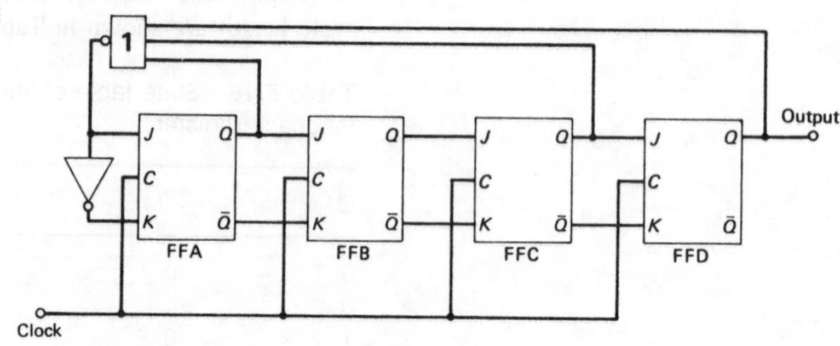

Fig. 5.46 Non-linear feedback shift register

Non-linear Feedback Shift Registers

If the feedback from the Q terminals of a shift register to its input terminal does *not* consist exclusively of exclusive-OR gates, the circuit is said to be non-linear.

The two possible states of a 4-stage Johnson counter are shown in Figs. 5.40(*a*) and (*b*). It is possible to move out of one cycle and into the other and back again at certain points. Figure 5.45 shows the two cycles combined, with connections linking 2 and 4, and 10 and 11. There are 16 different states and so any count up to 15 can be designed.

Suppose, for example, that a count of 7 using the sequence 0, 1, 2, 5, 10, 4, 8, 0 . . . is required. The state table is given by Table 5.18. From the table

$$J_A = \bar{Q}_A\bar{Q}_C\bar{Q}_D$$

The required circuit is shown in Fig. 5.46.

Table 5.18 State table of non-linear feedback register

State	Q_D	Q_C	Q_B	Q_A	Q_D^+	Q_C^+	Q_B^+	Q_A^+
0	0	0	0	0	0	0	0	1
1	0	0	0	1	0	0	1	0
2	0	0	1	0	0	1	0	1
5	0	1	0	1	1	0	1	0
10	1	0	1	0	0	1	0	0
4	0	1	0	0	1	0	0	0
8	1	0	0	0	0	0	0	0

Exercises 5

5.1 Two J-K master–slave flip-flops are connected to form a modulo 3 counter. Draw the circuit and, with the aid of a truth table and a timing diagram, explain its operation.

5.2 Show how 5 J-K flip-flops could be connected to form a twisted ring counter. What would be the count of this circuit? With the aid of a timing diagram explain its action.

5.3 Draw the diagram, and explain the action, of a shift register. Explain its operation when storing the binary number 0010.

5.4 Design a ripple counter whose outputs follow the 8421 numbering system.

5.5 Draw diagrams to show how a 4-bit shift register can be converted into *a)* a 4-bit ring counter, *b)* an 8-bit ring counter.

5.6 Design a divide-by-5 up-counter using a state table and map.

5.7 Draw the circuit of, and explain the operation of, a dynamic shift register. State some factors that affect the clock frequency.

5.8 Design a synchronous counter to pass through the following states in the order given: 011, 100, 101, and 111.

5.9 A linear feedback shift register has 4 stages *A, B, C* and *D*. Feedback paths exist between the outputs of stages *A* and *D* to the input of stage *A*. Determine the code cycle that is generated.

5.10 Design a counter with three stages to cycle through the states 0, 2, 1, 7, 5, 6, 0, ...

5.11 Draw the block diagram of a 3-stage shift register. Show the following inputs: serial data in, clock and reset.

5.12 Design a non-linear fsr to count through 11 states.

5.13 Design a 5-stage Johnson decade counter with decoding for alternate states starting with $Q_A = Q_B = Q_C = Q_D = Q_E = 0$.

5.14 Design a synchronous counter which counts from 0 to 14 inclusive and resets to 0.

5.15 Design a synchronous counter to count down from 9 to 0 then resets to 9.

5.16 Design a 4-bit counter to follow the sequence

3, 11, 0, 10, 9, 8, 4, 14, 15, 1, 7, 13, 5, 12, 6, 2, 3 . . .

5.17 Draw a diagram to show how a decade counter may be made using *a)* a 4-bit counter, *b)* a divide-by-5 counter and a flip-flop.

5.18 A non-synchronous counter is to count up to binary 1111 and then reset. Draw the block diagram of the circuit stating which type of flip-flops you are using.

5.19 *a)* Show how i) a D and ii) a J-K flip-flop can be connected as a divide-by-2 circuit. Explain the operation of the circuit with the aid of timing diagrams. *b)* Draw the diagram of a ripple counter with bcd outputs.

5.20 Determine the count sequence of the divide-by-9 circuit of Fig. 5.16. Note that it does not go from 0 to 8.

5.21 Show how two 7490 decade counters can be connected to give a count of 25.

6 Memories

Introduction

Many digital systems include some form of *memory* or *storage* where data can be held on either a permanent or a temporary basis. A number of magnetic devices, such as discs and tapes, are available for use as permanent stores and these possess the advantage of being *non-volatile*; this term means that they are able to retain stored data when the power supplies have been turned off. Semiconductor memories have now become the predominant memory technology, particularly for short-term storage, and their capacity is being increased yearly as improved techniques are developed. The reasons for the increasing use of integrated circuit memories are that they are smaller, cheaper, and faster to operate than the magnetic alternatives. They may be either volatile or non-volatile depending on the type.

A memory consists of a matrix of memory *cells* and a number of digital circuits that provide such functions as *address selection* and *control*. Each cell has a particular location in the matrix that is identified by a unique *address* and is able to store one bit of information. Two kinds of semiconductor memory are used, known as the *random access memory* or RAM, and the *read only memory* or ROM.

The basic requirements for both a ROM and a RAM are that

a) any location in the memory can be addressed;
b) data can be read out of an addressed location; and
c) (for a RAM only) data can be written into an addressed location.

The random access memory or RAM can have data read out of it, or written into it, with the same access time for all addresses. The *access time* is the time that elapses between the start of a memory request and the data becoming available.

Two other terms are also of some importance. The *read cycle time* is the minimum time that must elapse between successive read operations, and the *write cycle time* is the time that must elapse between successive write operations. In manufacturer's data sheets these parameters are usually given by timing diagrams (see Figs 6.1 and 6.2). The diagrams will illustrate the difference in meaning between *access time* and *read/write cycle time*.

Note that the address and data lines are shown as going both high and low simultaneously. The diagrams are always drawn in this way to indicate that the address lines, etc. may be either high or low according to the data on them.

The reasons for, and the function of, the chip select and the read/write lines are explained later in the chapter.

Fig. 6.1 Timing diagram for reading data from a RAM

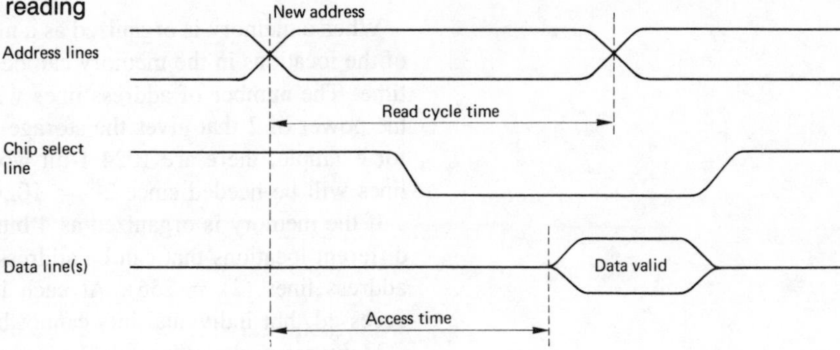

Fig. 6.2 Timing diagram for writing data into a RAM

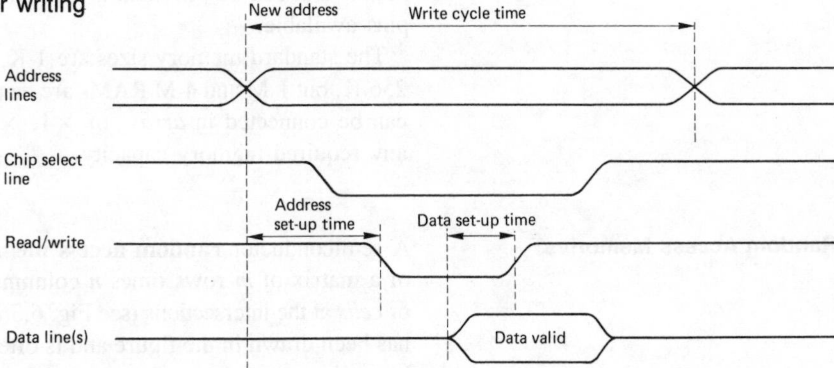

Both *dynamic* and *static* RAMs are possible, with the former being both cheaper and simpler to fabricate but possessing the disadvantage that the stored information must be periodically *refreshed*. On the other hand, the dynamic memory is faster and allows a greater packing density to be achieved.

The other kind of memory is the read only memory or ROM and it can only be used to store permanent data. Some types of ROM are programmed by the semiconductor manufacturer and can only be used to store the data designed into the device. Other types, known as *programmable read only memories* or PROMs, can be set up for a particular purpose after the equipment manufacturer has taken delivery of the memory.

The *capacity* of a memory is measured in terms of *words* and *bits*, with the total number of bits being quoted as so many K. The letter K does *not* denote 1000 but it represents 2^{10} or 1024. Thus a 1 K store is able to hold 1024 bits of information. This capacity could be organized in a number of different ways: for example, 256 4-bit words or 128 8-bit words or 1024 1-bit words.

When a memory is organized as a number of 1-bit words, any one of the locations in the memory can be individually addressed at any time. The number of address lines which are necessary is equal to the power of 2 that gives the storage capability of the memory. If, for example, there are 1024 1-bit word locations, then 10 address lines will be needed since $2^{10} = 1024$.

If the memory is organized as 4 bits per word, there will be 256 different locations that can be addressed and this will necessitate 8 address lines ($2^8 = 256$). At each location a 4-bit word will be accessed, but individual bits cannot be separately addressed.

Many memories store only 1-bit words because there is then only a need for *one* input/output data terminal. This is particularly true for the larger memories since the number of input and output terminals that can be provided is limited by the number of ic package pins available.

The standard memory sizes are 1 K, 4 K, 16 K, 32 K, 64 K and 256 K, but 1 M and 4 M RAMs are increasingly available. Memories can be connected in *arrays* of ×1, ×4, and ×8 to obtain almost any required memory capacity.

Random Access Memories

A semiconductor **random access memory** is arranged in the form of a matrix of m rows times n columns with the memory elements or *cells* at the intersections (see Fig. 6.3(a)). Although a square matrix has been drawn in the figure and is often used, e.g. 32 × 32 = 1024 bits, other matrices are also employed, e.g. 64 × 16 = 1024 bits. The cells are grouped to form words consisting, usually, of 1 bit, 4 bits or 8 bits per word. Each word has a unique address in the matrix and can be selected by the simultaneous application of the appropriate logic level to one row address line and one column (or group of columns) address line.

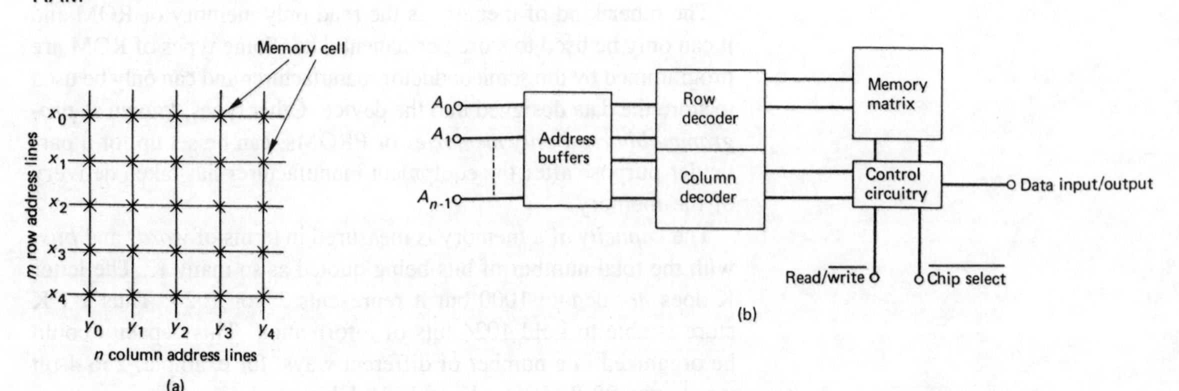

Fig. 6.3 (a) Basic arrangement of a RAM matrix, (b) block diagram of a RAM

To reduce the number of address input lines necessary, addresses are inputted in binary and decoded as shown by Fig. 6.3(*b*). The ic package must have the following pins:

Input and output data bits (separate or combined).
Address bits.
A read/write bit.
A chip select (or chip enable) bit.

The chip select or the chip enable input is used to select a particular memory from a number that are connected in parallel in order to produce a larger capacity memory. The data output lines are either open-collector (p. 33) or three-state. A *three-state output* can be placed into any one of three distinct states; it can be in either of the logical 0 or 1 states or it can be switched to have a very high output impedance. The third condition, in which the output is generally assumed to be open-circuited, is applied when the chip is disabled (chip select input at 1) and makes it possible for memory chips to be paralleled to form larger memories. (A three-state output is often known as *tri-state*.)

When the logical state of a location in a memory is to be *read*, the read/$\overline{\text{write}}$ line is set to the READ level (usually logical 1) and the address of the required location is fed in. The data held at that location is presented at the data output terminal(s) and there it remains until such time as any one, or more, of the address, read/write or enable inputs change state. The stored data is *not* affected by the read-out process; in other words the read-out is *non-destructive*.

When data is to be *written into* the memory, the read/write line is set to the WRITE level (usually logic 0). Then data present at the input data line(s) will be entered into the memory at the addressed location. Any data that was already at that location is *lost*. This does mean, of course, that the input data must not change whilst the chip is in the write state.

Organization of a RAM

As a simple example consider a 9-bit memory. This could be organized as three 3-bit words as shown by Fig. 6.4(*a*), or as nine 1-bit words as shown by Fig. 6.4(*b*). For the 3 × 3 memory the decoded address takes one of the X select lines high to select a particular word. All three cells in the same column have common data 0 and data 1 lines and the control circuitry is such that the input data D_{in} is written into the memory when the read/$\overline{\text{write}}$ line goes high. Conversely, when the read/$\overline{\text{write}}$ line is low, the addressed word is read out of the memory.

The 9×1 memory is organized differently (Fig. 6.4(*b*)). Each cell is addressed in two planes and so each cell can be individually located. The data 0 and data 1 lines are fed to each cell in parallel as shown and, since only one bit at a time can be read out of, or written into, the memory, only one control circuit block is shown.

Some specific examples of decoding for RAMs follow:

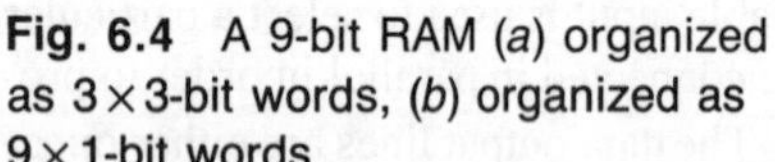

Fig. 6.4 A 9-bit RAM (*a*) organized as 3×3-bit words, (*b*) organized as 9×1-bit words

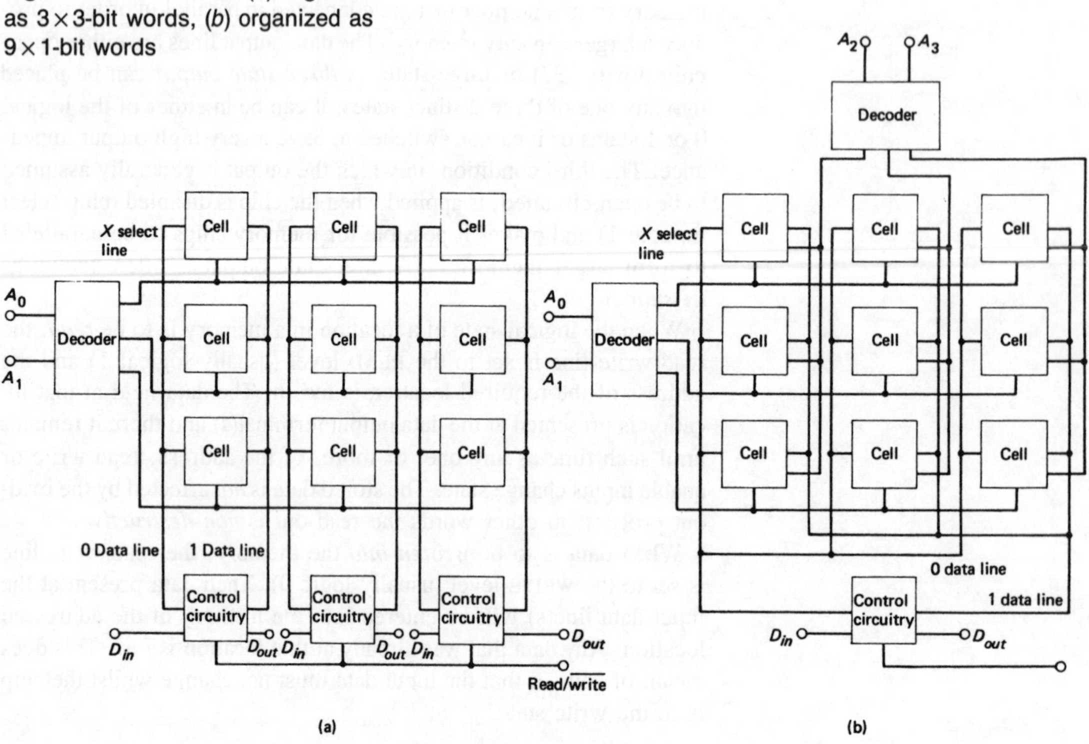

Fig. 6.5 1 K × 1 memory matrix

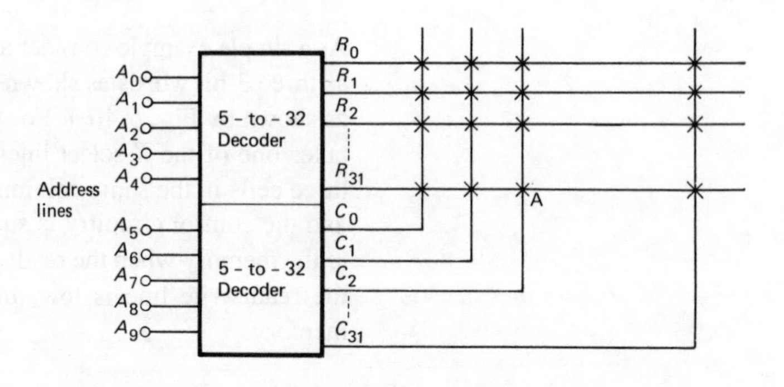

1 1 K × 1 using a 32 × 32 matrix
The memory matrix consists of 32 rows and 32 columns with a memory cell at each intersection. The row and column decoders must be able to select any one out of 32 lines and hence they each require 5 address lines ($2^5 = 32$). The arrangement of the matrix is shown by Fig. 6.5.

If the input address is 0001011111, the addressed cell will be at location $R_{31}C_2$ (marked as A). The number of address pins necessary is 10 and the number of data pins will be either 1 or 2.

Fig. 6.6 256 × 4 memory matrix

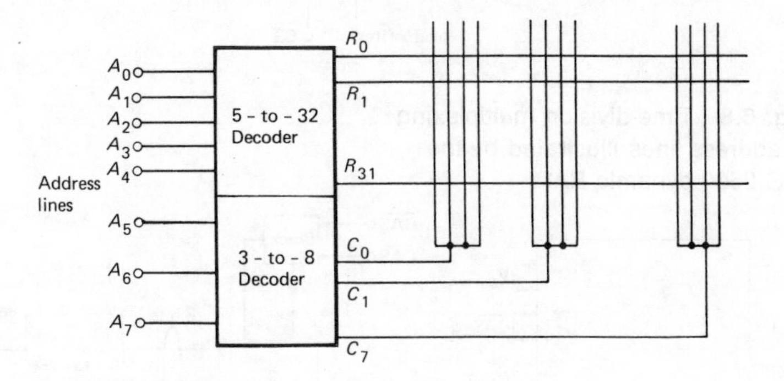

2 256 × 4 using a 32 × 32 matrix
The arrangement of the 256 × 4 RAM is shown by Fig. 6.6. Columns are now grouped in fours to produce 4-bit words and hence there are 8 groups of columns. The column decoder has to be able to select any 1 from 8 column groups and so it must have 3 address lines ($2^3 = 8$). The number of data pins required is either 4 or 8.

The input address 001 00010 would select the 4-bit word stored at the location in row R_2 group C_1.

Two further examples are given in Figs 6.7(*a*) and (*b*). These show how the same pins can be used for both input and output data under the control of the read/$\overline{\text{write}}$ input signal. The number of pins needed for an ic package accommodating the RAM of Fig. 6.7(*a*) would be 14 address pins, 4 data input/output pins, 1 read/$\overline{\text{write}}$ pin, 1 chip select pin, and two pins for $+V_{CC}$ and earth. This is a total of 22 pins so that a 24-pin package would probably be employed. Clearly, the number of pins required for a 64 K memory organized as 64 K × 1-bit words would be even larger because of the increase in the number of address pins necessary.

For this reason some of the larger memories employ time-division multiplexing of the address lines. One example of this technique is shown in Fig. 6.8 for the IMS2600 64 K × 1 dynamic RAM. The matrix is arranged as 256 rows × 256 columns and the address inputs are time-division multiplexed under the control of the RAS and CAS clocks. Normally, $\overline{\text{CAS}}$ is high when $\overline{\text{RAS}}$ goes low. Then the 8

Fig. 6.7 (*a*) 16 K × 4 memory matrix, (*b*) 8 K × 8 memory matrix

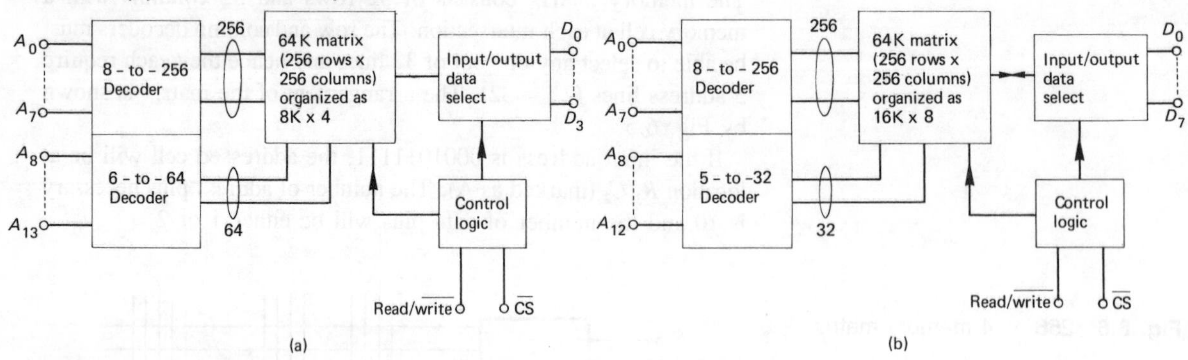

Fig. 6.8 Time-division multiplexing of address lines illustrated by the IMS 2600 dynamic RAM

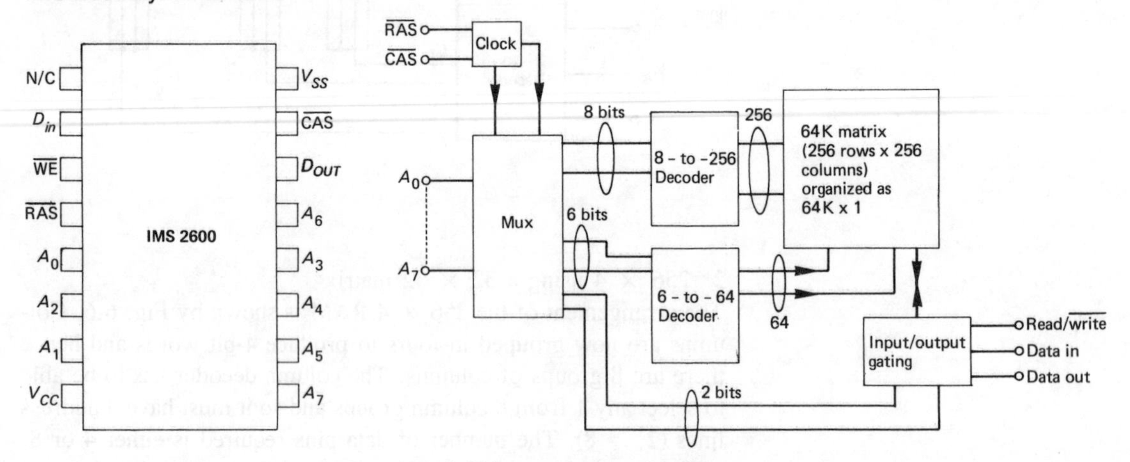

address inputs are latched and decoded to select one of the 256 rows in the matrix. This row address must be held for a minimum period and then it can be switched to the wanted column address. Once the column address is stable $\overline{CAS}$ can be taken low and then the column address is latched and used to select the wanted column. The address cycle ends when $\overline{RAS}$ goes high.

Expanding Memory Capacity

Very often the number of words to be stored and/or the number of bits per word is larger than the capacity of a single memory chip. A larger capacity memory can be constructed by the suitable combination of two or more chips.

A 1 K RAM may be arranged to provide a 1 K × 1 bit memory. This means that it will have ten address lines ($2^{10} = 1024$), one data input, one data output, one read/$\overline{write}$ line, and one chip select line.

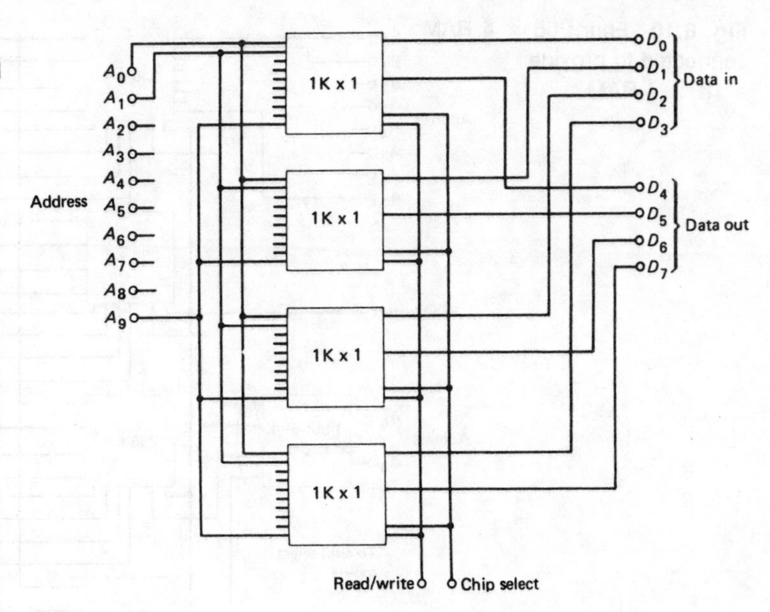

Fig. 6.9 Four 1K × 1 rams connected to provide a 1K × 4 RAM

Suppose that a 1 K × 4-bit memory is required. Four 1 K memory chips are necessary and must be connected with their address pins in parallel. The read/$\overline{\text{write}}$ pins and the chip select pins must also be connected in parallel. The necessary pin connections are shown in Fig. 6.9. When the chips are enabled, the same location is selected in each memory and the data can be read out of pins D_4, D_5, D_6 and D_7 or, if the WRITE mode has been selected, written into pins D_0, D_1, D_2 and D_3.

Modules are available which combine a number of memory chips in this manner. One such module provides a 64 K × 8-bit dynamic RAM consisting of eight 65 K × 1 RAMs mounted upon a substrate. Another module combines four 16 K × 4 RAMs to give a 16 K × 16 memory.

The number of words that a memory can store can be increased by combining two or more chips in the manner shown in Fig. 6.10. Here four 256 × 4-bit (again 1 K) chips are connected with their address, read/$\overline{\text{write}}$, and input/output data lines in parallel. The chip select inputs of the four chips are connected to the four outputs of a 1-out-of-4 decoder.

The address lines A_0 through to A_7 select the required memory cell in each of the four memory matrices but *only one* of the chips is enabled by the 2-bit address A_8/A_9 applied to the decoder. Thus, one only address at a time in the combined memory can be read out of or written into.

Modules are also offered for this aspect of memory expansion. One module comprises four 65 K × 1 RAM connected to give a 256 K × 1 memory.

Fig. 6.10 Four 256 × 4 RAM connected to provide a 1K × 4 RAM

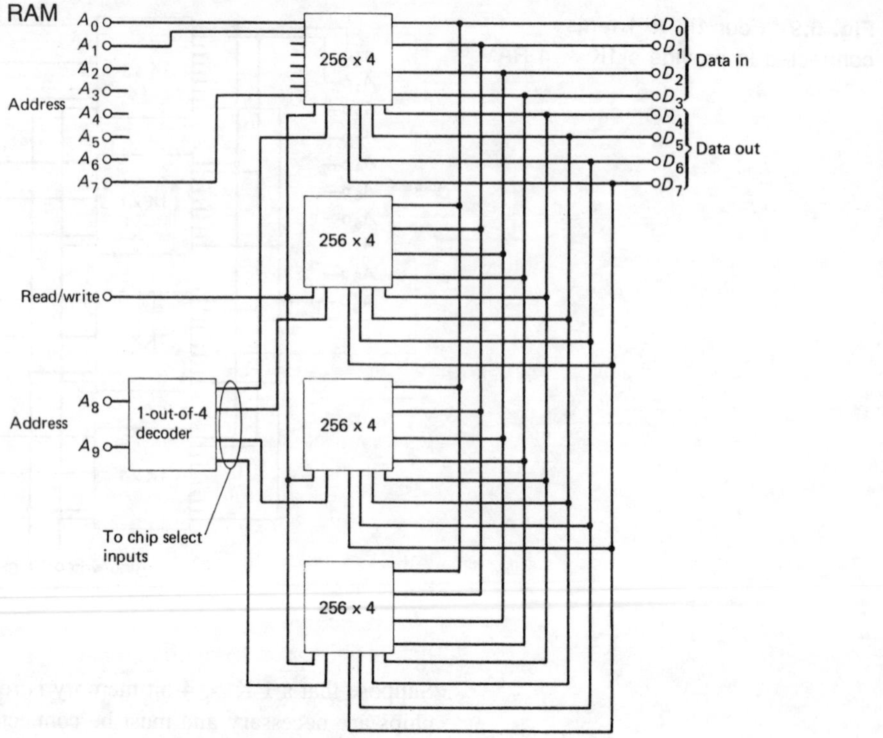

Static Random Access Memories

There are two main classes of RAM manufactured, namely **static RAMs** and **dynamic** RAMs. Static RAMs can be manufactured using either bipolar transistor or mosfet technology but only mosfet techniques can be employed in the manufacture of a dynamic RAM.

A **static RAM** consists of a matrix of flip-flops each of which is able to store one bit of information and acts as a memory cell. Bipolar transistor RAMs are mainly in the ttl family although ecl devices are also available, while mosfet devices are either nmos or cmos.

The circuit of a *ttl memory cell* is shown in Fig 6.11. Transistors T_1 and T_2 are connected as a flip-flop. The collector supply voltage is shown as +3.5 V but may be any value up to +5 V. This means that logic 0 is represented by voltages below about +0.3 V and logic 1 by voltages above about +3 V. When the cell is in the state T_1 ON and T_2 OFF, the circuit stores a 0 bit. Conversely, a 1 bit is stored by the state T_1 OFF and T_2 ON. An n-p-n transistor with multiple emitters will turn OFF when *all* its emitters are held at a voltage that is more positive than the base potential. If any one or more of the emitters are at approximately 0 V, and the base is positive, the transistor will conduct no matter how positive the other emitters should be.

Fig. 6.11 TTL memory cell

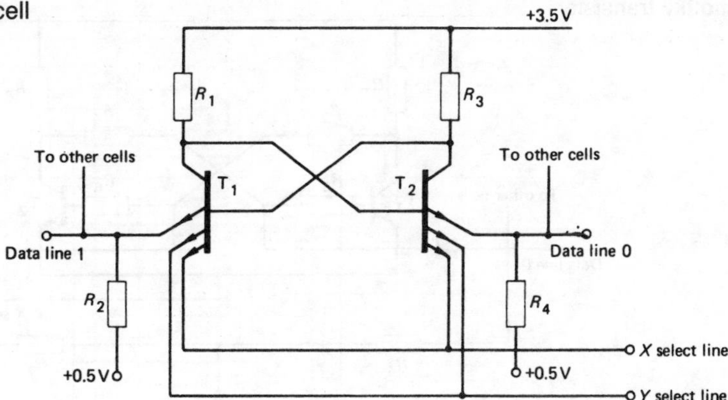

Suppose that both the X and Y lines are at logic 0 and that the logic state of the cell is T_1 ON and T_2 OFF. The current conducted by T_1 will flow out of the two emitters connected to the X and the Y lines but no current flows into the data 1 line because this emitter is held at a voltage greater than the 0 logic level. If *either* the X *or* the Y line is addressed so that its voltage rises to the logic 1 level, there will be no change in the operation of the circuit since one emitter of each transistor will still be low.

When the cell is addressed by *both* the X and Y lines having a logic 1 voltage level applied, the associated emitters become more positive than the "data" emitter. The current conducted by the ON transistor now flows out of the cell into the data line. This current then flows into circuitry not shown in the figure and it causes the data output terminal to assume the same logic level as the cell.

When data is to be written into an addressed cell, either the data 1 line or the data 0 line is taken to the logic 1 voltage level to turn OFF the associated transistor (all three emitters are then high). A cell that has not been addressed will not pass current to the appropriate data line and does not respond to the write operation because each transistor will have at least one emitter at the logic 1 voltage level.

The circuit of the *Schottky transistor memory* cell is shown in Fig. 6.12. When the cell is not addressed, line X is at the logic 1 voltage level or at about +2.5 V. This means that the two diodes D_1 and D_2 are reverse-biased and so non-conducting. Suppose that the flip-flop is in the state T_1 ON and T_2 OFF and that the cell is addressed by the voltage on line X being reduced to the logic 0 level of about +0.3 V. The base potential of T_1 will then be equal to $(+0.3 + V_{BE})$ volts $\simeq$ 1.0 V and the diode D_1 will become forward biased by approximately 0.5 V. The voltage on the data 0 line will fall from +1.5 V to 1.0 + 0.4 = 1.4 V since the voltage drop across a Schottky diode is about 0.4 V. The collector potential of T_1 is now equal to 0.3 V + $V_{CE(sat)} \simeq$ 0.5 V and so the voltage on the data 1 line is much greater than the voltage change on the data 0 line. The *difference*

Fig. 6.12 Schottky transistor memory cell

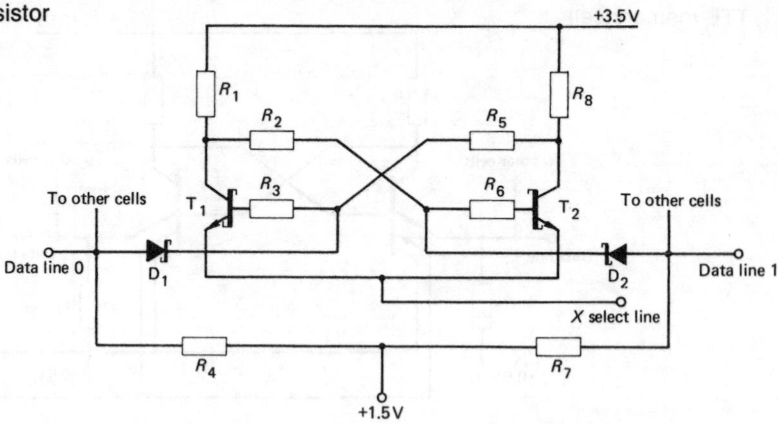

between the two voltage changes is used as an indication of the logical state of the cell. Only this cell is read since all the other cells will have their *X* line high so that their diodes D_1 and D_2 are OFF.

When data is to be written into a cell, the voltage of the X line is reduced to 0.3 V to turn either D_1 or D_2 ON. Suppose that as before the cell is in the state T_1 ON and T_2 OFF and that the other state is required. The voltage of the data 1 line is then increased to about +2.8 V to turn transistor T_2 ON. The normal flip-flop action then ensures that T_1 rapidly turns OFF. If the required state had been T_1 ON and T_2 OFF, the +2.8 V voltage would have been applied to the data 0 line and this would, of course, have had no effect on the logical state of the cell. Non-addressed cells are not written into since their *X* line voltage remains at +2.5 V and so the collector potential of their ON transistor is $2.5+0.2=+2.7$ V. Thus a data line voltage of $+2.7+0.4=+3.1$ V is needed to initiate switching.

The Schottky transistor memory cell possesses advantages over the multiple-emitter transistor cell in that it is faster to operate and it dissipates less power. On the other hand it can be addressed in the *X* plane only.

Some of the memories in the ttl family are: 7489/74170/74LS170 (4×4), and the 74LS670 (4×4). The first two have a totem pole output, while the last RAM has a three-state output. TTL memories are of relatively small capacity, and are mainly employed as fast, small temporary stores.

Mosfet Static Cells

When larger storage capacities are needed a mosfet circuit must be used. This is because mosfet circuitry allows much greater packing densities to be achieved so that large capacity memories can be fabricated within a single chip. An added advantage is that the power

Fig. 6.13 NMOS memory cell

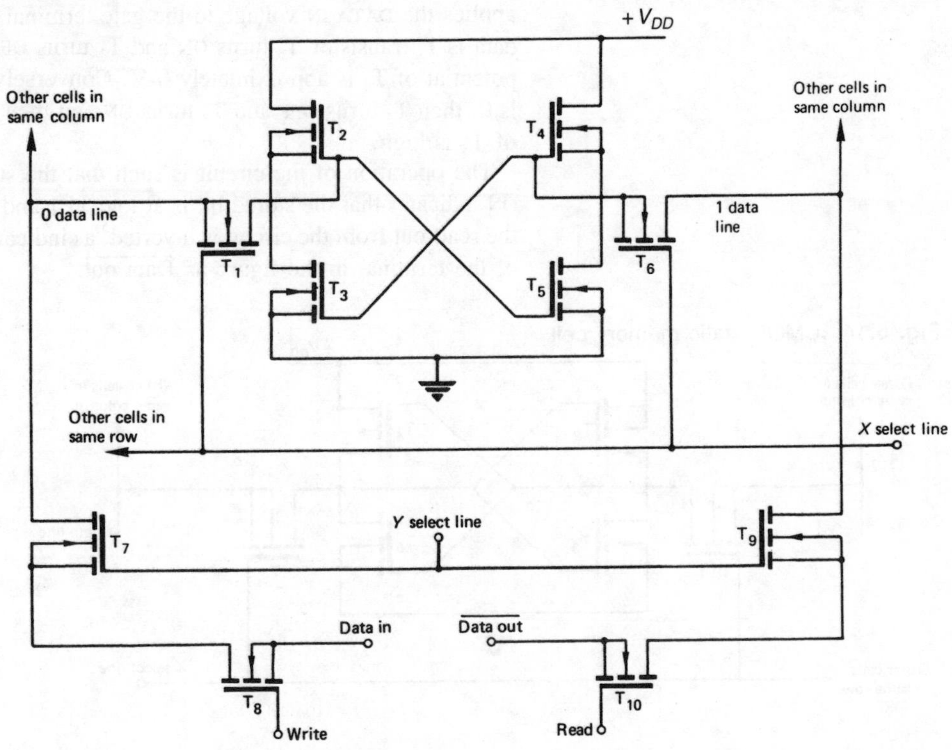

consumption of a mosfet cell is much less than that of a bipolar transistor cell. The price paid for these advantages is a reduced speed of operation.

The circuit of an *nmos memory cell* is shown in Fig. 6.13. Transistors T_3 and T_5 are connected as a flip-flop with active drain load resistances provided by T_2 and T_4.

Transistors T_1 and T_6 form gates which open to connect the cell to the data lines when the cell is addressed. Six transistors form one memory cell. Transistors T_7, T_8, T_9 and T_{10} form a part of the control circuitry and are common to all the cells in the same column of the memory matrix. Note that all the transistors are n-channel enhancement-mode mosfets that are able to act as bi-directional switches.

The memory cell is addressed by making both the *X* and the *Y* select lines go high (approximately $+V_{DD}$ volts). The *X* select voltage turns transistors T_1 and T_6 ON and connects the cell to the 0 and 1 data lines. The *Y* select voltage turns T_7 and T_9 ON to connect the write and read circuits to the selected column and hence to the addressed cell.

To read the state of a cell, the READ terminal is taken high, turning the transistor T_{10} ON. The drain potential of T_5 then appears, with little voltage drop, across T_6, T_9 and T_{10}, at the data-out terminal. For new data to be written into the cell, the WRITE terminal is set to the logical 1 voltage level. This turns ON transistor T_9 and applies the DATA IN voltage to the gate terminal of T_5. If the input data is 1, transistor T_5 turns ON and T_3 turns OFF so that the drain potential of T_5 is approximately 0 V. Conversely, if the input data is 0, then T_5 turns OFF and T_3 turns ON and then the drain potential of T_5 is high.

The operation of the circuit is such that the state T_3 OFF and T_5 ON indicates that the stored bit is at logical 1 and vice versa. Hence the read-out from the circuit is inverted, as indicated by the labelling of the terminal in the figure as $\overline{\text{Data out}}$.

Fig. 6.14 CMOS static memory cell

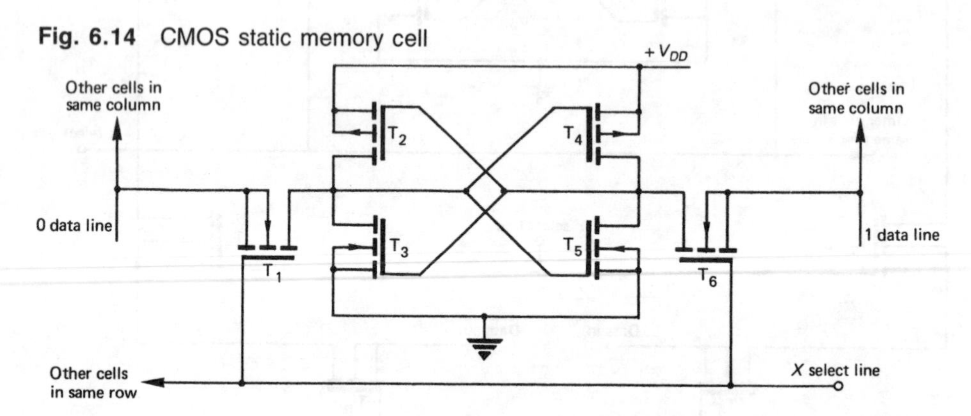

Fig. 6.14 shows the circuit of a *cmos memory cell*. The pairs of transistors forming the opposite sides of the flip-flop consist of one p-channel and one n-channel mosfet. The use of complementary mosfets in this way increases the speed of operation and reduces the power dissipation. As in the nmos circuit, mosfets T_1 and T_6 connect the cell to the data lines when the *X* select line is high. The bit stored is 1 when the circuit is in the state T_3 OFF and T_4 ON and 0 when it is T_3 ON and T_4 OFF.

MOS memories are of larger capacity than their bipolar transistor alternatives and an example, with pin connections, is shown in Fig. 6.15. Other memory sizes are given in Table 6.1. CMOS memories have hitherto been considerably slower to access than other types of memory but this situation is changing with the introduction of high-speed cmos logic. Modern cmos memories are now very nearly comparable in speed with nmos and they have possess a considerable advantage in lower power consumption. However, hcmos is, at present, more expensive than nmos.

Some memories employ a combination of both cmos and nmos technology to provide low-power devices with nmos speed of operation. This technique is known as merged mos; it is also used for some dynamic RAMs. With such circuits, the memory cells utilize nmos techniques but the decoder and other circuitry employ cmos logic.

Cmos RAMs can be made virtually non-volatile by the provision

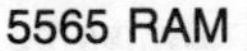

Fig. 6.15 Pin connections of the TC 5565 RAM

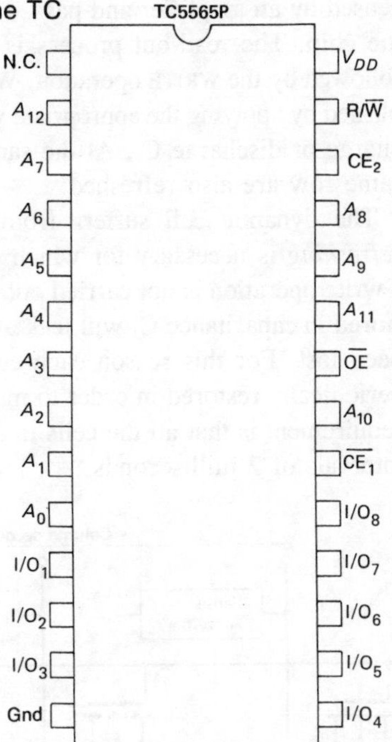

Pin names	
A_0 to A_{12}	Address
$R/\overline{W}$	Read/Write control input
$\overline{OE}$	Output Enable input
$\overline{CE_1}$, CE_2	Chip Enable inputs
I/O_1 to I/O_8	Data Input/Output
V_{DD}	Power +5V
Gnd	Ground
N.C.	No Connection

of a self-contained battery supply that can be switched into circuit should the power supply fail. Since cmos devices dissipate very little power only a small capacity battery is required, say 0.4 Ah.

Dynamic Random Access Memories

A reduction in the number of transistors needed per cell can be achieved if a **dynamic memory** is employed. In this type of memory data is stored in the stray capacitances which exist between the gate and the source terminals of a mosfet. This allows a larger number of cells to be fabricated within a given chip area. Dynamic RAMs offer the lowest cost per bit of any type of memory and for this reason they are commonly adopted for large memory systems.

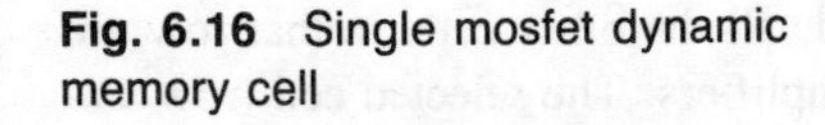

Fig. 6.16 Single mosfet dynamic memory cell

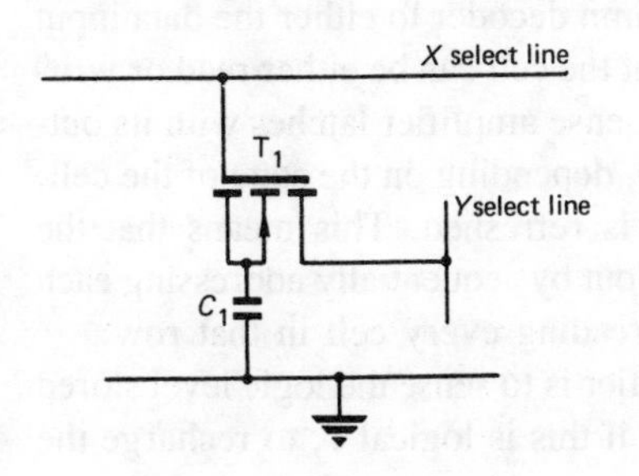

The early dynamic RAMs employed memory cells which used four mosfets but further development has led to the single-mosfet cell, the basic circuit of which is shown by Fig. 6.16. The cell consists of a capacitance C_1 in which data is stored by the absence or presence of a charge. The capacitance is accessed through a single gating transistor T_1.

Zero charge corresponds to the logic 0 state and full charge to the logic 1 state. To read the state of the cell, it must first be addressed by applying a voltage to the *X* select line to turn T_1 ON. The charge, or lack of charge, on C_1 is then fed to the *Y* access line and is there sensed by an amplifier and passed on to the data output terminal of the chip. The read-out process is destructive and must always be followed by the WRITE operation. Writing a bit into the cell is accomplished by applying the appropriate voltage to the *Y* select line to either charge or discharge C_1. At the same time all the other cells in the same row are also refreshed.

The dynamic cell suffers from the disadvantage that periodic *refreshing* is necessary for which extra logic must be provided. If a write operation is not carried out within a certain time, any charge stored in capacitance C_1 will leak away and the stored data will have been lost. For this reason each cell must have its charge (if any) periodically restored in order to maintain the stored data. The usual requirement is that all the cells in a dynamic RAM are refreshed at intervals of 2 milliseconds.

Fig. 6.17 Organization of a dynamic RAM

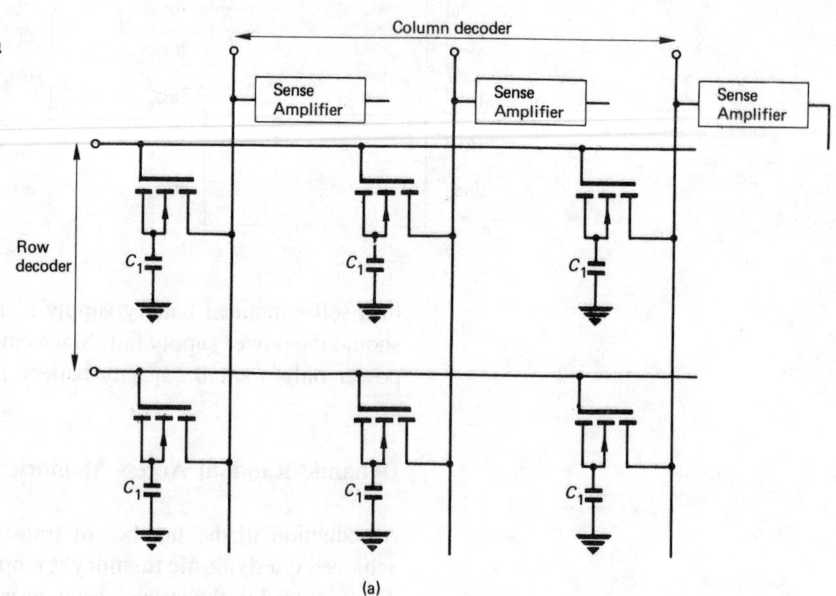

Figures 6.17(a) and (b) show the way in which dynamic RAM cells are arranged in a number of rows and columns. Whenever a cell in any particular row is addressed, all the other cells in that row are also connected to their sense amplifiers. The selected cell (row *and* column) is connected via the column decoder to either the data input or the data output terminals so that the cell can be either read or written into. When a cell is read, the sense amplifier latches with its output at either logical 1 or logical 0, depending on the state of the cell, and if it is 1 the stored charge is refreshed. This means that the refreshing process can be carried out by sequentially addressing each row in the memory matrix and reading every cell in that row.

The function of the sense amplifier is to sense the logic level stored in a connected memory cell and, if this is logical 1, to recharge the

Fig. 6.17 continued

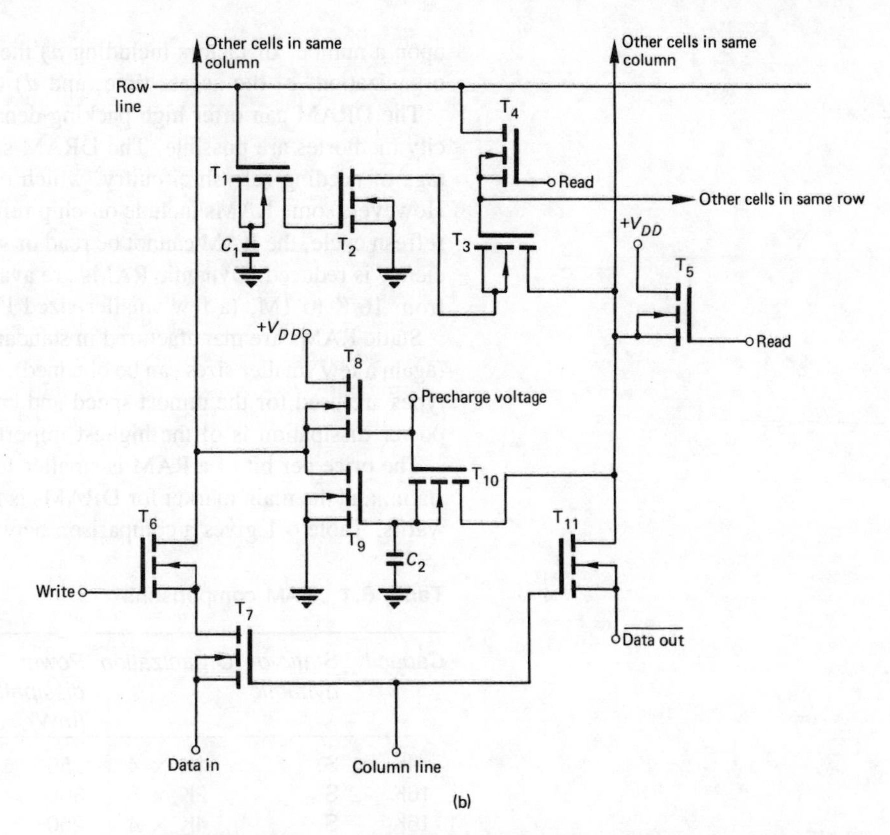

(b)

storage capacitor. The way in which this is carried out by the sense amplifier is as follows. The capacitance C_1 associated with the sense amplifier is charged to the minimum voltage that is recognized as representing logical 1 level. The sense amplifier then compares the voltages across C_1 and C_2. If the voltage across C_1 is greater than the voltage across C_2, the output of the sense amplifier latches to the logical 1 voltage level and this recharges the capacitance C_1 to refresh the stored charge. Conversely, if the voltage across C_2 is less than the voltage across C_1, the output of the sense amplifier latches low and the capacitance C_1 is not recharged.

The number of address pins on a modern dynamic RAM is one-half of the total number needed, since time-division multiplexing of the address pins with two address fields is employed. The first of these is known as the row address field and the second is called the column address field.

The industry standard dynamic RAM is 64 K × 1 but other sizes are also available (see Table 6.1).

Comparison between Static and Dynamic RAM

The choice of a suitable RAM for a particular application depends upon a number of factors including *a*) the total bit capacity, *b*) the organization, *c*) the access time, and *d*) the power consumption.

The DRAM can offer high packing density so that high bit capacity memories are possible. The DRAM suffers from the disadvantage of needing refresh circuitry, which often requires extra logic. However, some RAMs include on-chip refresh circuitry. During the refresh cycle, the RAM cannot be read or written into so that its efficiency is reduced. Dynamic RAMs are available in standard sizes of from 16 K to 1M, (a few smaller-sized DRAMs do also exist).

Static RAMs are manufactured in standard sizes of 4 K up to 64 K (again a few smaller sizes can be obtained). Generally speaking, nmos types are used for the utmost speed and cmos types are used if low power dissipation is of the highest importance.

The price per bit of a RAM is smaller for dynamic types than for static, and the main market for DRAMs is for capacities of 16 K upwards. Table 6.1 gives a comparison between some typical RAMs.

Table 6.1 RAM comparisons

Capacity	*Static or dynamic*	*Organization*	*Power dissipation (mW)*	*Access time (ns)*
4K	S	1K × 4	250	150
16K	S	2K × 8	300	100
16K	S	4K × 4	250	150
64K	S	8K × 8	200	130
16K	D	16K × 1	450	225
64K	D	64K × 1	300	100
64K	D	16K × 4	275	150
256K	D	256K × 1	350	150

Read Only Memories

Read only memories or **ROMs** have data permanently written into them either by the manufacturer or by the user, and can *only* be used to read out data. Data is *not* written into the memory during a circuit operation. Read only memories are used for a variety of purposes including storage of mathematical tables (such as logarithms and trigonometric functions) and code conversion and program storage. A ROM is non-volatile.

A ROM is organized in a similar way to a RAM in that data is stored at different locations in the memory matrix and each location has a unique address. When a particular location is addressed, the data stored at that address is read out of the memory. The read-out is non-destructive. Address decoders are used to reduce the number of input address bits required and so allow the ROM to be fabricated within a standard dil package.

Figure 6.18(*a*) shows the arrangement of a diode ROM. Although diodes are shown, bipolar or field effect transistors could be used instead, connected in the manner shown by Figs. 6.18(*b*) and (*c*). In any case, the ROM is an integrated circuit device fabricated within an ic package. The logical output of the ROM is determined by the connections made by the diodes between an address decoder line, 0 through to 7, and an output line, *A* through to *F*. Normally, each output decoder line is high and goes low only when it is selected by an input address. Each diode that is connected to a low decoder output line is turned ON and takes the associated output line low also.

If, for example, the input address is 010, decoder output line 2 will go low. This row line has two diodes connected to it; these two diodes conduct and take the output lines *A* and *E* low also. Thus the output of the ROM is 011101. If the input address is 110, decoder line 5 goes low and the ROM output is 110110. The output signal will remain available for as long as the address is held.

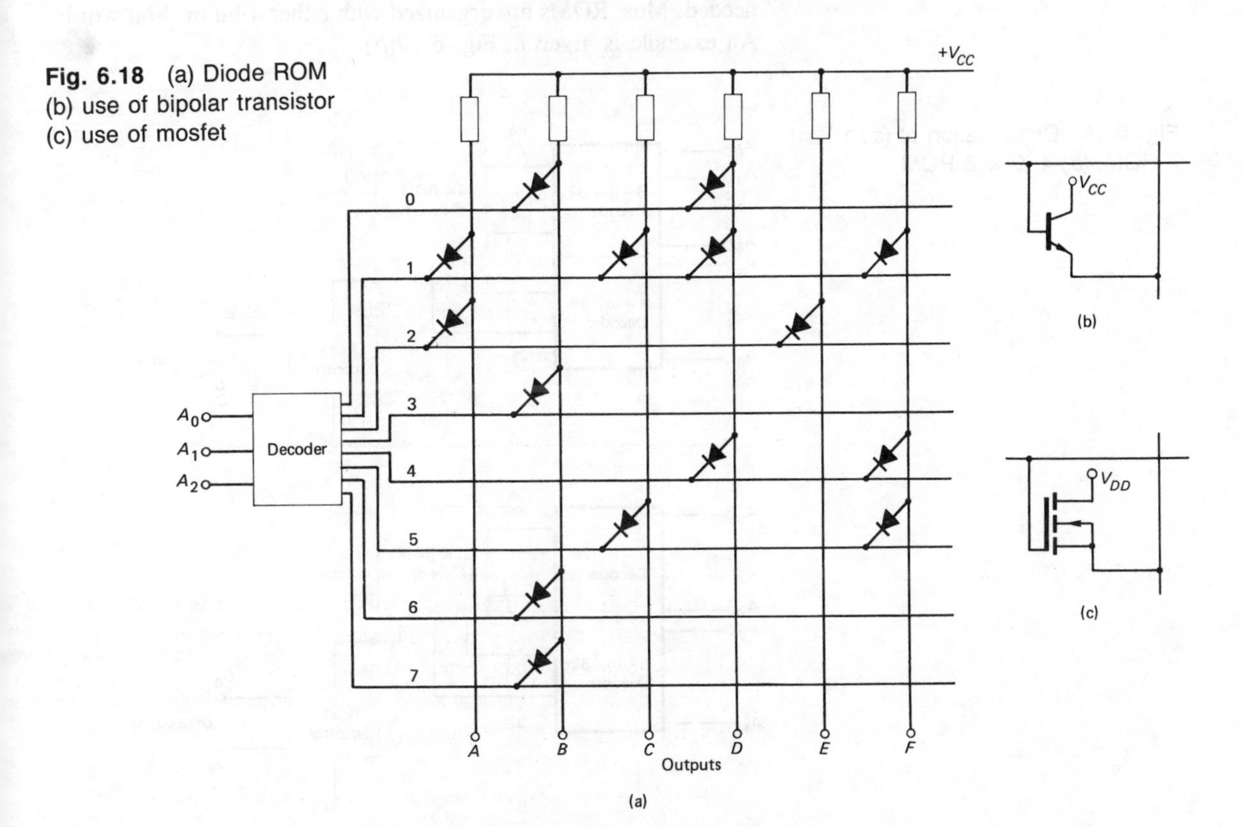

Fig. 6.18 (a) Diode ROM (b) use of bipolar transistor (c) use of mosfet

When either kind of transistor is used as the conducting element, a decoder output going high will turn ON the associated transistors and give the logical 0 state at each of the connected outputs. Those outputs that are not linked by a transistor to the selected row will remain at logical 1.

When data is stored in a ROM the number of ROM outputs should correspond to the number of bits per word. If the arrangement shown in Fig. 6.18(*a*) is used as it stands this would not be the case, unless, of course, the number of columns was only four, or perhaps eight. A typical size for a ROM is 4 kilobits and, if this is organized with a square matrix, there will be 64 rows and 64 columns.

Suppose that the 4 K bit memory is organized as 1028 4-bit words. The row address decoder must have six input bits since $2^6 = 64$. The four data output bits, D_0, D_1, D_2 and D_3 must be obtained from 64 columns. Thus, the columns will be divided into 16 groups of four bits, and this means that there is a need for a column decoder also. Since $16 = 2^4$ the column circuit must have four input bits to address the required groups of columns in the matrix. The required circuit is shown by Fig. 6.19(*a*).

It should be evident that if the matrix is not square, and/or the number of bits per word is different, other decoder circuits will be needed. Most ROMs are organized with either 4-bit or 8-bit words. An example is given in Fig. 6.19(*b*).

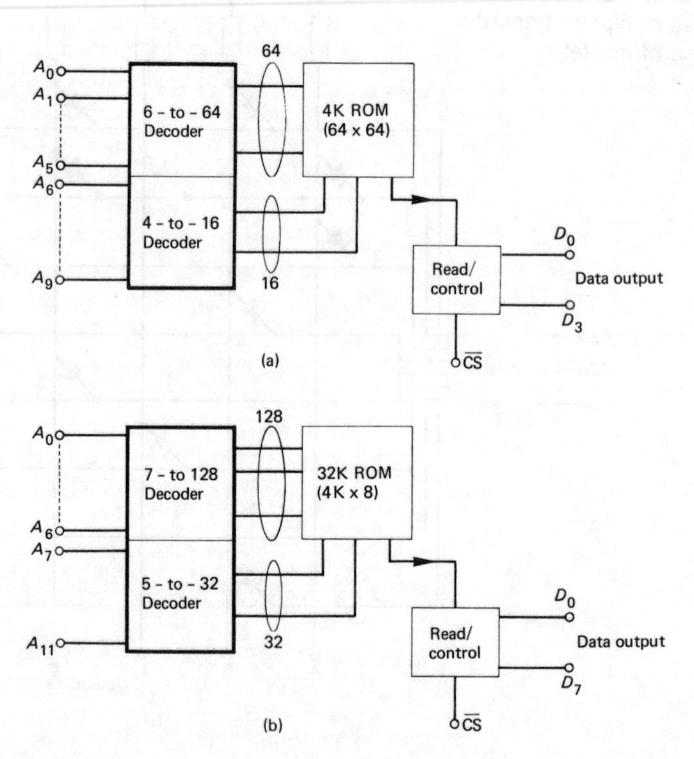

Fig. 6.19 Organization of (*a*) 1 K × 4 ROM, (*b*) 4 K × 8 ROM

Expansion of the capacity of a ROM can be accomplished by combining chips in similar manner to that described earlier for the RAM.

Commercially available ROMs include the following: 2048 × 8-bit,

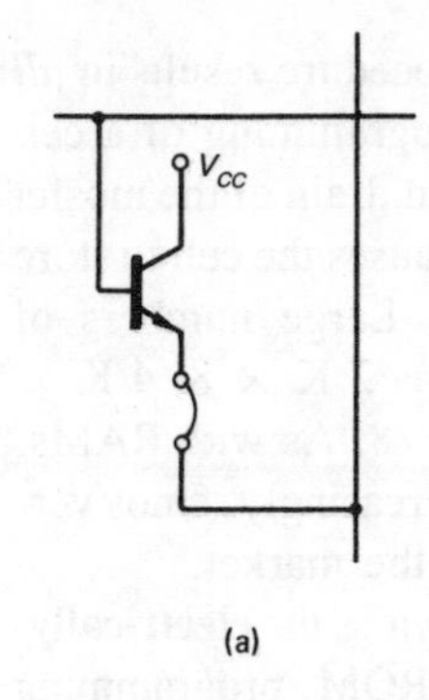

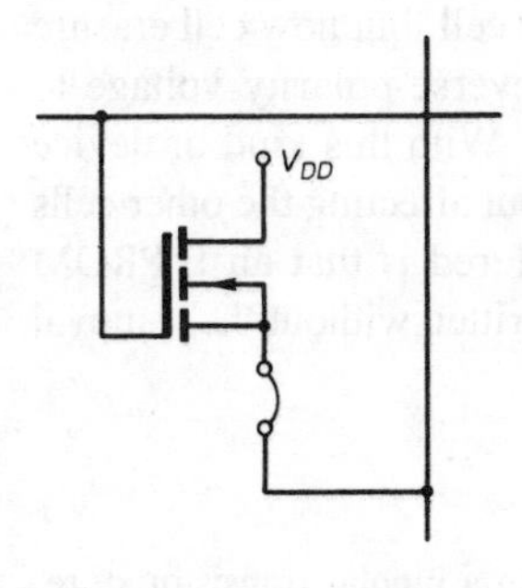

Fig. 6.20 Elements in (*a*) a ttl prom, (*b*) a mosfet prom

16 K × 8-bit, 256 × 4-bit, 256 × 8-bit, 1024 × 4-bit and 1024 × 8-bit.

Access times are generally in the region of 60 ns to 250 ns.

A ROM is programmed for a particular purpose during the manufacturing process and its function cannot be altered by the user. This can be a useful feature for an equipment manufacturer who may have a need for a large number of identical devices. However, it is often much less convenient for the user of only small quantities of ROMs. In order to provide some flexibility in the possible applications for ROMs, *programmable* devices have been introduced.

Programmable ROMs

A **programmable ROM** or **PROM** is manufactured as a generalized integrated circuit with *all* the matrix intersections linked by *fusable* diodes or transistors (Fig. 6.20). Because the circuit *is* generalized it can be mass-produced, which, of course, brings the cost down. This means that *all* outputs will go to logical 0 whenever *any* of the decoder outputs is addressed. The purchaser of a PROM can program the device to perform for a specific application.

Programming is carried out by addressing a particular intersection by selection of its row and its column and then taking the base, or the gate, of the transistor to a high positive voltage and the column line to near earth potential. The current taken by the transistor is then large enough to blow the "fuse" and leave the emitter, or the source, of the transistor open-circuited. Each link thus destroyed will result in the associated output line remaining at logical 1 when that location is addressed.

The programmer must be very careful when carrying out this work since any links that may be incorrectly fused cannot be restored.

PROMs are widely used in the control of electric equipment such as washing machines and electric ovens. Bipolar PROMs are faster and cheaper than mosfet PROMs but their power dissipation is higher. Also, mosfet PROMs can have much larger matrices.

Erasable and Electrically-alterable PROMs

A further development of the PROM has led to the introduction of devices that can be reprogrammed if necessary. These devices utilize a different mosfet cell structure in which programming is accomplished by the storage of a charge in the cell instead of by the blowing of a fuse. With an **erasable PROM** or **EPROM**, erasure of an existing program is achieved by exposing the chip to ultraviolet radiation directed through a window in the chip package. The radiation increases the conductivity of the cell and in so doing allows the stored charge to leak away. The process cannot be applied to an individual cell but only to the whole matrix, and so the erasure procedure results in *all* the stored data being lost. After erasure, reprogramming of a cell is carried out by applying voltages to the gate and drain of the mosfet and earthing the source and the substrate. This causes the cell to store the charge that indicates the logical 1 state. Large numbers of EPROMs are produced in standard sizes such as 2 K × 8, 4 K × 8, 8 K × 8, 16 K × 8, 32 K × 8 and 64 K × 8. As with RAMs, power dissipation is an important factor and, increasingly, cmos versions of EPROMs are being introduced on to the market.

Another kind of reprogrammable ROM is known as the **electrically-alterable PROM** or **E^2PROM**. As with the EPROM, programming a cell to store a 1 bit is done by charging the cell, but now cell erasure is accomplished by the application of a reverse-polarity voltage to the mosfet that removes the stored charge. With this kind of device individual cells can be reprogrammed without affecting the other cells in the matrix. An important advantage offered is that an E^2PROM can be totally reprogrammed, or partly rewritten without the removal of the device from the circuit.

Exercises

6.1 Sketch and describe the circuit of a cell used in a bipolar transistor store. Draw a circuit to show how the cells are connected to each other to form a matrix.

With the aid of a block diagram explain how a particular location in the store is addressed.

6.2 Distinguish between coincident selection and linear selection of a memory cell and give some advantages and disadvantages of each method.

Draw the circuit of a coincident selection RAM cell and explain its operation.

6.3 Design a ROM to read out the squares of the numbers 1 through to 7.

6.4 Explain, with the aid of a suitable diagram, the operation of any kind of ROM. Explain the difference between a ROM and a PROM. What are the advantages of the latter?

What is an E^2PROM and when would it be used?

6.5 Draw the circuit diagram of a dynamic RAM cell and briefly explain its operation. Why is periodic refreshing necessary? Describe how particular locations in a 16 K RAM can be addressed if the organisation is *a*) ×1 and *b*) ×4.

6.6 Show how a number of 16 K × 1 memories can be combined to produce a 48 K × 1 memory. Design the necessary decoder.

6.7 Figure 6.21 shows the pin connections of the 74LS89 64-bit RAM that is arranged internally as a 16 × 4-bit word. Draw a diagram to show how two such circuits can be combined to make a 16 × 8-bit memory. State, without drawing a circuit, how a 32 × 4-bit memory could be achieved.

Fig. 6.21

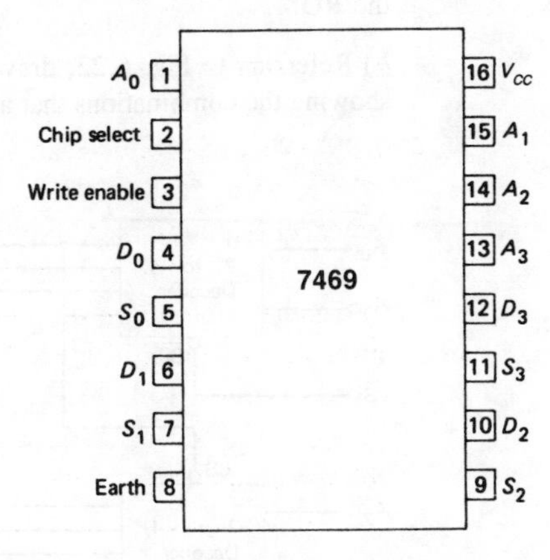

A = address, D = data in, S = sense (data) out

6.8 Explain the difference between dynamic and static memories. Draw the circuit, and explain the operation, of a cell in each.

6.9 *a*) Explain clearly the differences between a RAM and a ROM. State when each would be used.

b) Explain what is meant by the terms access time and cycle time when applied to a semiconductor memory.

c) List the relative advantages and disadvantages of (i) bipolar and mosfet static RAMs, (ii) static and dynamic RAMs. Why are bipolar transistor dynamic RAMs not available?

6.10 A 256-bit memory is arranged as 32 words of 4 bits each. How many input/output pins are needed on the dil package? Given that standard packages have an even number of pins, suggest a more efficient memory size from a pin utilization point of view. Would this memory be a static or a dynamic type and why?

6.11 How many 7489 16 × 4-bit memory chips must be combined to produce (i) a 64 × 4-bit memory, (ii) a 64 × 8-bit memory? Give the block diagram of each arrangement and outline its operation.

6.12 The description of a commercially available RAM is: it is a 64-bit RAM organized as a 16 word × 4-bit array; address inputs are buffered and fully decoded on chip; the outputs are three-state and are high impedance when the $\overline{\text{CS}}$ input is high.

Explain briefly the meaning of each sentence.

6.13 *a*) A ROM chip has 10 address pins and 8 data pins. State the capacity of the ROM.

b) Referring to Fig. 6.22, draw the truth table for address inputs A_0 to A_3 showing the combinations that allow memory chips C_1 to C_7 to be selected.

Fig. 6.22

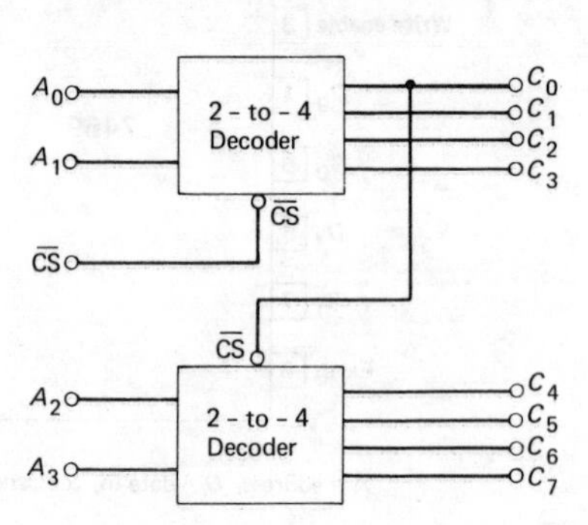

7 Digital-to-Analogue and Analogue-to-Digital Converters

Introduction

There are many areas in electrical/electronic/communication engineering where analogue techniques still represent the best solution to a problem. The parameters which occur in practice and that may need to be processed in some way are all analogue in nature. Examples are many and include pressure, flow, position and temperature. For digital processing or measurement, these parameters need to be first converted into analogue electrical signals, then changed into digital form, processed, and then returned to their original analogue form.

Fig. 7.1 Principle of a digital-to-analogue converter

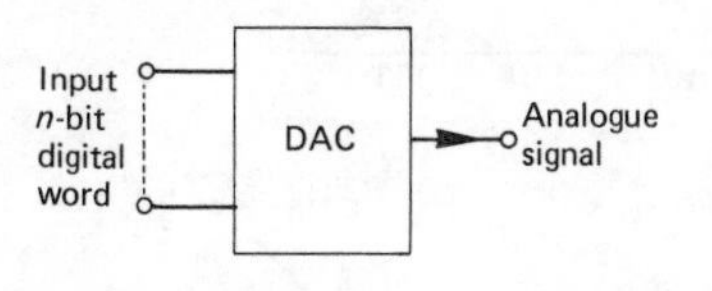

The function of a digital-to-analogue converter (dac) is to convert an input digital word into the corresponding analogue voltage. The principle of operation of a dac is shown by Fig. 7.1. Conversely, the function of an analogue-to-digital converter (adc) is to convert an input analogue voltage into the equivalent digital word (see Fig. 7.2).

Fig. 7.2 Principle of an analogue-to-digital converter

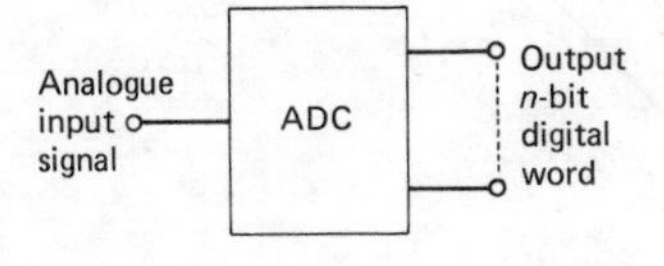

Digital-to-analogue conversion is a simpler technique than analogue-to-digital conversion; indeed, many adcs include a dac as one element in their circuit. For this reason, the concept of, and typical circuits for, digital-to-analogue conversion will be discussed before those of analogue-to-digital conversion. Also, circuits known as sample-and-hold amplifiers are often employed in conjunction with adcs to store analogue information during conversion. Sample-and-hold amplifiers will also be discussed before analogue-to-digital converters.

Digital-to-Analogue Converters

A **digital-to-analogue converter** (dac) is a circuit that accepts a digital signal at its input terminals and then generates an equivalent analogue signal at its output terminals. Essentially, a dac consists of a voltage or current reference, a number of binary weighted resistors, electronic switches, and some means of summing a number of weighted currents. Most commonly the input digital word uses the natural binary code and Fig. 7.3 shows the ideal transfer function for a 3-bit dac.

For each word, a particular analogue output voltage is generated. The ideal characteristic is also known as the *gain curve* and it joins the points 0 and the full-scale range (FSR). The practical characteristic is a stepped approximation to Fig. 7.3 and this means that some error inevitably exists. The error can always be reduced by increasing the number of input bits and thereby increasing the number of possible

Fig. 7.3 Transfer function of a dac

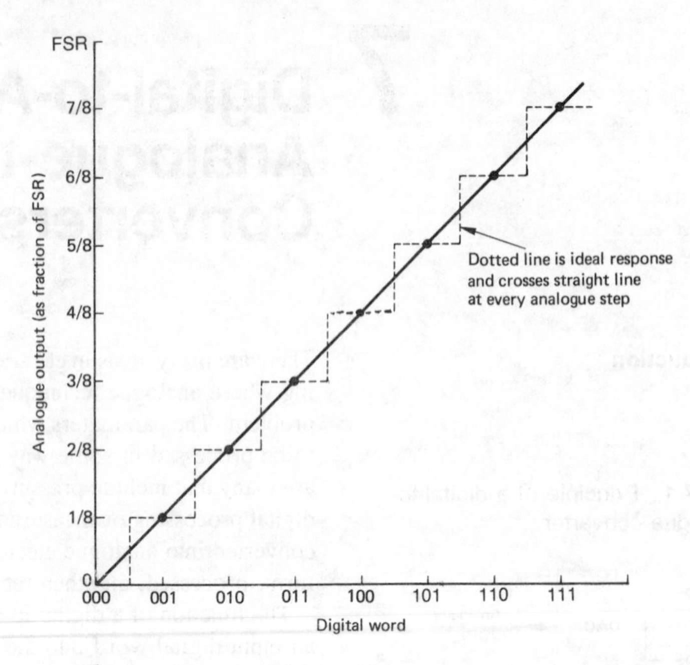

Fig. 7.4 Transfer function of a 4-bit dac

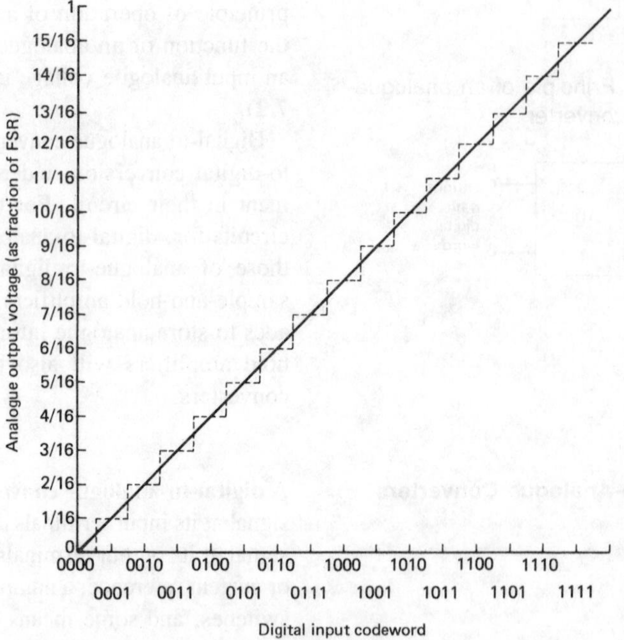

analogue output voltages. When three input bits are used, only 2^3 (or 8) different analogue voltages can be signalled by the circuit; if the number of bits is increased to 4, this becomes 2^4 (or 16) different voltages, and so on. This is illustrated by the transfer function given in Fig. 7.4. The error will be even smaller if 8 or 12 bits are employed or, put another way, the *resolution* of the circuit will be improved.

The value marked as FSR in both Fig. 7.3 and Fig. 7.4 is the *full-scale range* of the dac. This is the value which is divided into 2^n parts to determine the least significant bit. The analogue output of the dac can never quite attain the FSR since its maximum value is always *one* bit less. The least significant bit is equal to FSR/2^n and the most significant bit is equal to FSR/2.

Example 7.1

The full-scale range of a dac is 10 V. If the dac is *a*) a 3-bit circuit, *b*) an 8-bit circuit, calculate the magnitude of the output voltage represented by i) the lsb, ii) the msb, and iii) the FS.

Solution

a) i) lsb $= 10/2^3 = 1.25$ V
ii) msb $= 10/2 = 5$ V
iii) FS $= 10 - 1.25 = 8.75$ V

b) i) lsb $= 10/2^8 = 10/256 = 39.1$ mV
ii) msb $= 10/2 = 5$ V
iii) FS $= 10 - 0.0391 = 9.609$ V

The FS is the full-scale output of the dac and it is the maximum output voltage it can deliver to a load. The FS is produced, for natural binary, when *all* of the input bits are at logical 1.

For an *n*-bit dac, the FS is given by

$$(2^n - 1)/(2^n \times \text{FSR})$$

Two other codes are employed for the input digital word of a dac, both of which are bipolar. The first of these is the *offset binary code* and it is obtained by offsetting the natural binary code so that the half-scale word becomes 0 V. The transfer characteristic for a dac that uses the offset binary code is shown by Fig. 7.5(*a*). The other binary code is the *twos complement code* and this is obtained by inverting the msb of the offset binary code. The transfer characteristic of a dac using the twos complement code is given in Fig. 7.5(*b*).

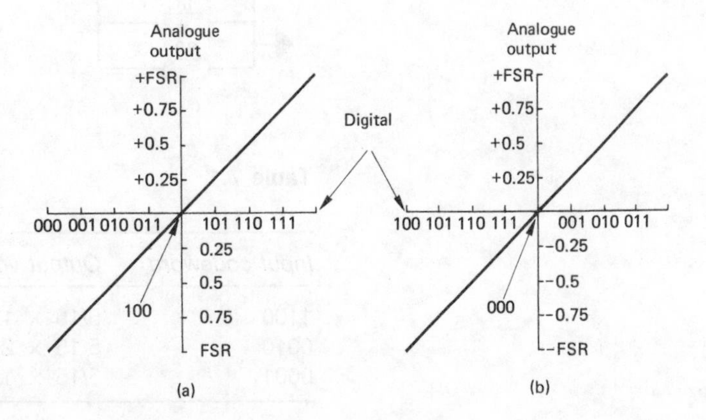

Fig. 7.5 (*a*) Use of offset binary code, (*b*) use of twos complement code

Digital-to-Analogue Converter Circuits

Most dacs are parallel circuits in which all of the bits of the input digital word change state simultaneously. A relatively few dacs are serially operated and these have the input bits applied to the circuit sequentially.

Figure 7.6 shows four resistors connected to the outputs of a 4-bit parallel output shift register and having their other terminals connected in common. The resistor values bear a binary relationship to one another as shown.

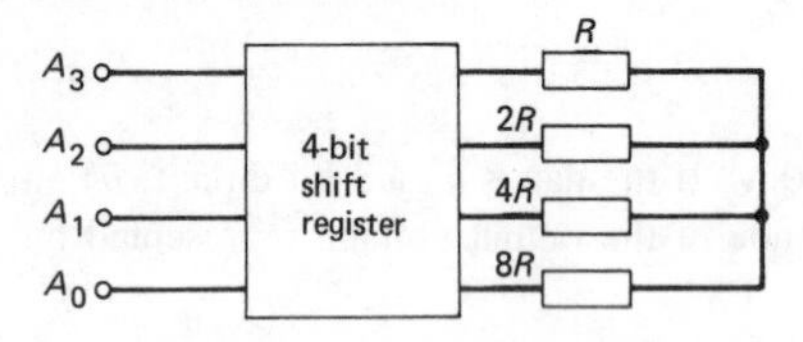

Fig. 7.6 Basic dac

The voltage applied to a resistor by its associated register output is either 0 V or +5 V, depending on whether that output is low or high. Suppose, for example, that the input digital word is 1000. Then the circuit can be redrawn as shown by Fig. 7.7(*a*); this, in turn, can be redrawn as shown by Fig. 7.7(*b*). From Fig. 7.7(*b*) the output voltage of the circuit is equal to

$$\frac{5 \times 8R/7}{R + 8R/7} = \frac{5 \times 8}{15} \text{ volts}$$

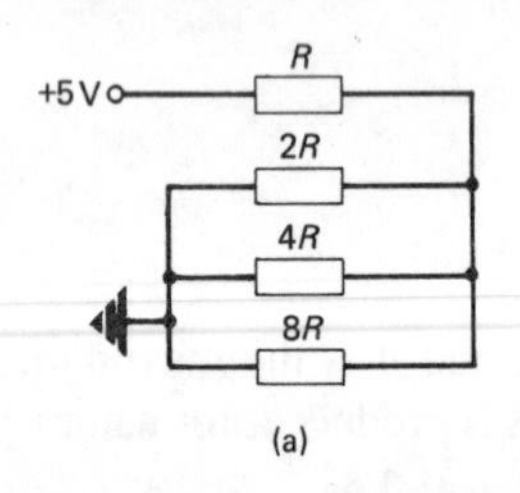

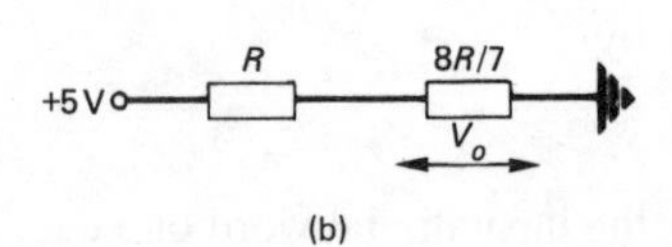

Fig. 7.7 Figure 7.6 redrawn for input codeword of 1000

Similarly, when the input digital word is 0100, the circuit can be redrawn as shown by Fig. 7.8(*a*) and (then) (*b*). From Fig. 7.8(*b*)

$$V_0 = \frac{5 \times 8R/11}{2R + 8R/11} = \frac{5 \times 8}{30} = \frac{5 \times 4}{15} \text{ volts}$$

It should be noted that when the input signal is 1000 the analogue output voltage is 5/15 × 8 volts, and when the input word is 0100 the output voltage is 5/15 × 4 volts. Similar results are obtained for other input digital words and a few examples are given in Table 7.1. Clearly, the output voltage is directly proportional to the decimal number represented by the input codeword.

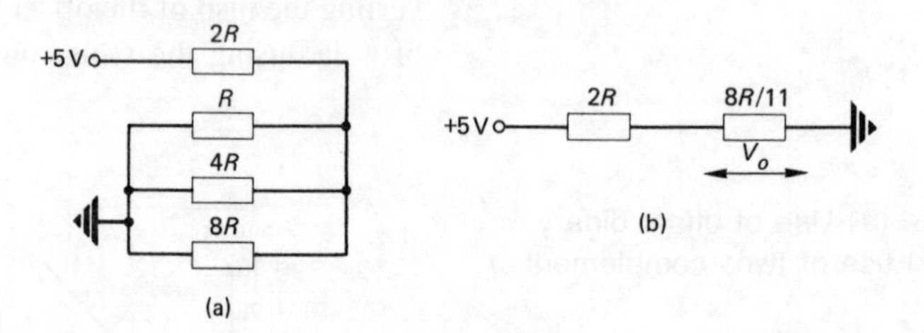

Fig. 7.8 Figure 7.6 redrawn for input codeword of 0100

Table 7.1

Input codeword	*Output voltage*	*Decimal*
1100	5/15 × 12	12
0010	5/15 × 2	2
0001	5/15 × 1	1

Fig. 7.9 Weighted resistor dac

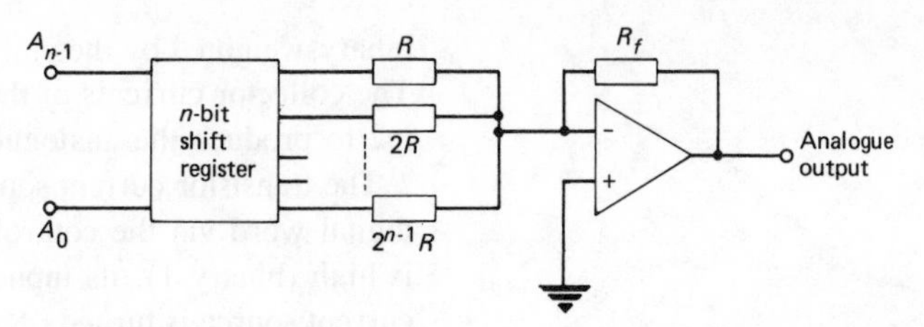

The voltage V_o appearing at the output of the circuit is generally applied to the input of an op-amp as shown by Fig. 7.9.

Example 7.2

Calculate the output voltage of the circuit given in Fig. 7.9 if $R = R_f$ and the digital input word is *a*) 1010 and *b*) 0110. Suppose that each register output produces a voltage of either 0 V or +5 V.

Solution

$$a)\quad V = -5R\left(\frac{1}{R} + \frac{0}{2R} + \frac{1}{4R} + \frac{0}{8R}\right)$$

$$= -5(1 + 1/4) = -6.25 \text{ V} \quad (Ans)$$

$$b)\quad V = -5R\left(\frac{0}{R} + \frac{1}{2R} + \frac{1}{4R} + \frac{0}{8R}\right)$$

$$= -5(1/2 + 1/4) = -3.75 \text{ V} \quad (Ans)$$

A variation of the basic circuit just described is illustrated by the *weighted current source* circuit given in Fig. 7.10. Essentially, the circuit consists of an array of switched current sources which are

Fig. 7.10 Weighted current source dac

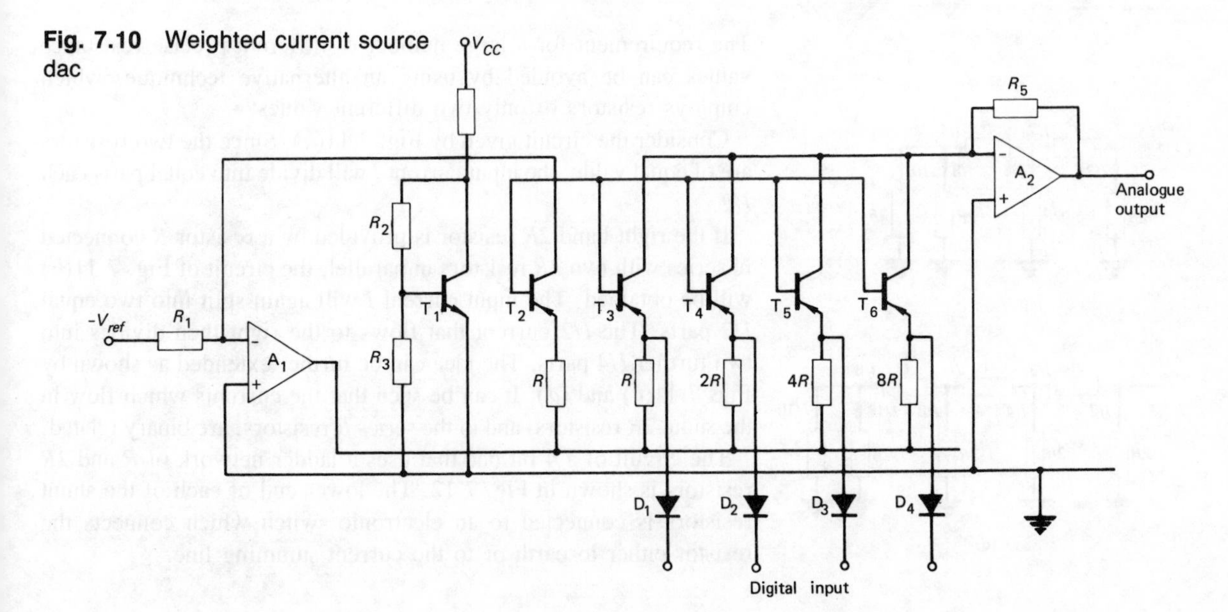

binary-weighted by the emitter resistors of value R, $2R$, $4R$ and $8R$. The collector currents of the transistors are summed by the op-amp ic2 to produce the analogue output voltage.

The transistor current sources are switched ON or OFF by the input digital word via the control diodes D_1 to D_4. When a digital input is high (binary 1), its input diode is turned OFF and the associated current source is turned ON. Conversely, if a digital input is low, its diode conducts and the current supplied by the op-amp ic1 flows via the diode. Therefore, the associated current source will be OFF.

The base voltages of each of the current sources are set by the T_1, R_1, R_2, R_3 circuit and the emitter currents are maintained at a *constant* value by the control amplifier ic1 and its associated voltage reference together with T_2.

The currents produced by each of the ON current sources are summed by the op-amp ic2 and converted into an analogue output voltage. Some version of this technique is employed in several integrated designs of dac.

A difficulty associated with the weighted resistor technique of digital-to-analogue conversion is the need for a number of precise-valued resistors, particularly if several bits are used. Suppose that a 12-bit converter is required and that the most significant resistor has a value of 1000 ohms. The other resistors should then have *precise* values of 2 kΩ, 4 kΩ, 8 kΩ, . . . , 2^{12-1} kΩ or 2.048 MΩ. It is difficult, and hence expensive, to produce a number of precise-valued resistors over such a wide bandwidth especially if the circuit design is to be integrated.

Fig. 7.11

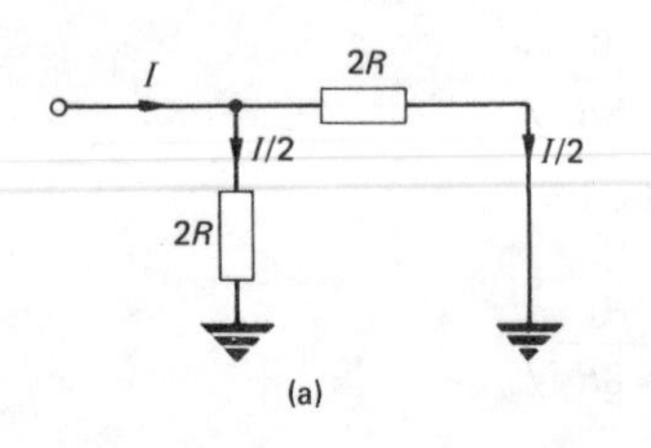

(a)

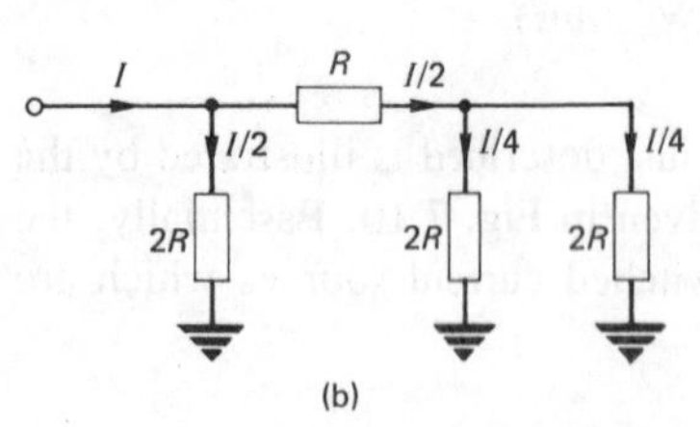

(b)

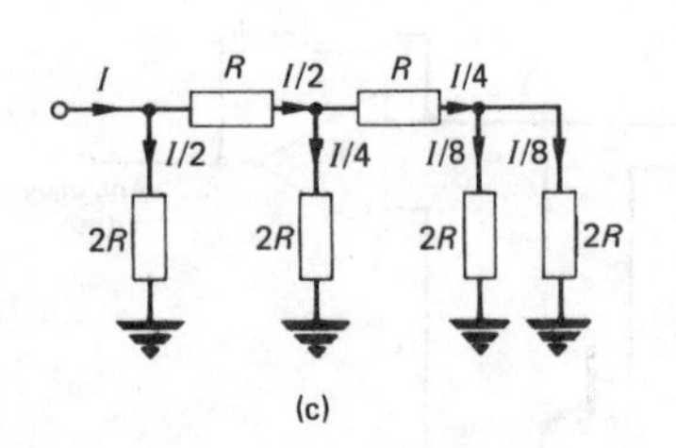

(c)

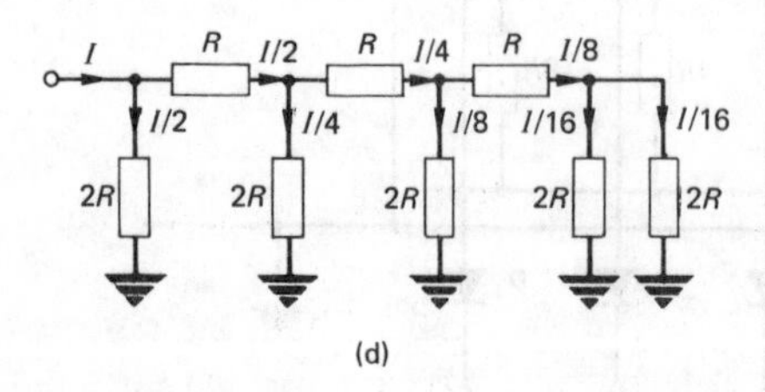

(d)

R-2R Circuit

The requirement for a large number of different precise resistance values can be avoided by using an alternative technique, which employs resistors of only two different values.

Consider the circuit given by Fig. 7.11(*a*). Since the two resistors are of equal value, the input current I will divide into equal parts each $I/2$.

If the right-hand $2R$ resistor is provided by a resistor R connected in series with two $2R$ resistors in parallel, the circuit of Fig. 7.11(*b*) will be obtained. The input current I will again split into two equal $I/2$ parts. The $I/2$ current that flows to the right then divides into two further $I/4$ parts. The idea can be further extended as shown by Figs 7.11(*c*) and (*d*). It can be seen that the currents which flow in the shunt $2R$ resistors, and in the series R resistors, are binary related.

The circuit of a 4-bit dac that uses a ladder network of R and $2R$ resistors is shown in Fig. 7.12. The lower end of each of the shunt resistors is connected to an electronic switch which connects the resistor either to earth or to the current-summing line.

The input digital word is applied to the shift register whose outputs operate the electronic switches. The most significant bit switches a current of $I/2$, the next most significant bit switches a current of $I/4$, and the least significant bit switches a current of $I/16$. The currents switched to the line connected to the op-amp are summed and converted into the wanted analogue output voltage.

Example 7.3

Calculate the output voltage of the circuit given in Fig. 7.12 when the input digital word is 1111, if V_{ref} = 5 V, R_f = 10 kΩ and R = 5 kΩ.

Solution

Current into ladder network is

$$\frac{V}{R_{in}} = \frac{V}{R} = \frac{5}{5\ \text{k}\Omega} = 1\ \text{mA}$$

Fig. 7.12 R-2R dac

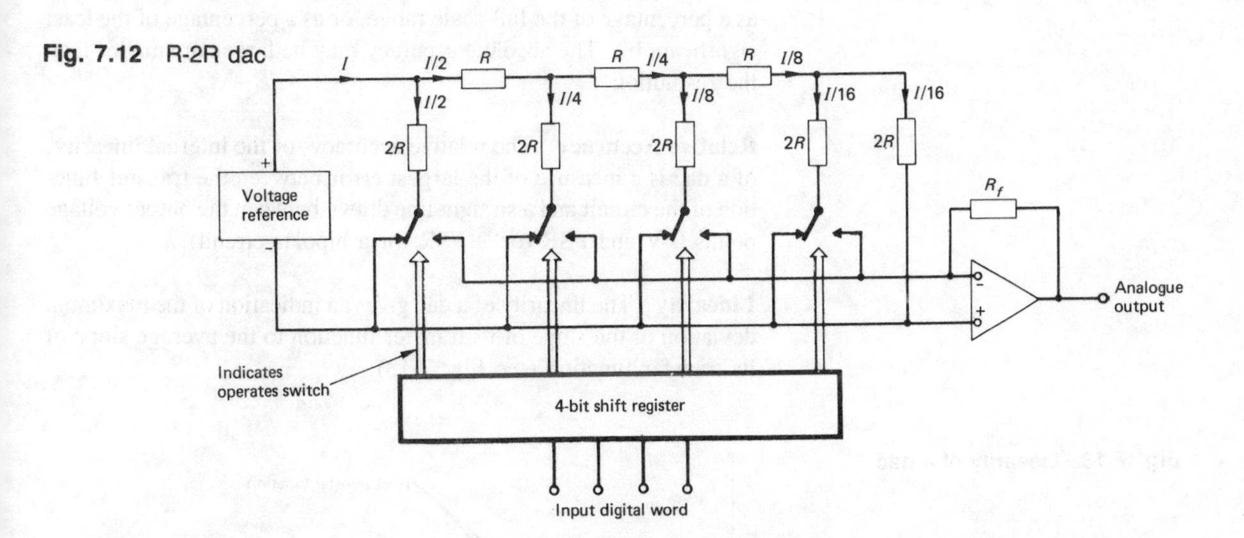

a) When the digital input is 1111, all currents are switched to the op-amp. Hence, the summed current is

$$I/2 + I/4 + I/8 + I/16 = 15I/16 \quad \text{or } 15/16\ \text{mA}$$

Therefore

$$V = \frac{-15}{16} \times 10^{-3} \times 10^{4} = \frac{-75}{8} = -9.375\ \text{V} \quad (Ans)$$

The advantage of the R-2R technique is the requirement for only two values of precise resistance. There are some potential sources of error including *a*) the accuracy and stability of both the voltage reference source and the resistors, and *b*) the accuracy of the current-to-voltage conversion of the op-amp.

Terms used with Digital-to-Analogue Converters

There are a number of terms commonly employed to describe the performance of a dac.

Resolution The resolution of a dac is determined by the number of bits in the input digital word since this sets the number of possible analogue output voltages. If there are *n* bits, the resolution is equal to $1/2^n$. If, for example, an 8-bit dac is considered, the number of possible output voltages is 2^8 (or 256). This means that the smallest change in the output voltage is 1/256 times the full-scale range of the circuit. The resolution of the circuit is 1/256 (or 0.4%).

Absolute Accuracy The absolute accuracy of a dac is the largest difference between the actual output voltage and the output voltage predicted by the ideal transfer characteristic when a given input digital word is applied to the circuit. The absolute accuracy can be quoted as a percentage of the full-scale range, or as a percentage of the least significant bit. The absolute accuracy may be larger or smaller than the resolution.

Relative Accuracy The relative accuracy, or the internal linearity, of a dac is a measure of the largest error between the transfer function of the circuit and a straight line drawn between the output voltage points 0 V and FSR (or ±FSR for a bipolar circuit).

Linearity The linearity of a dac gives an indication of the maximum deviation of the slope of its transfer function to the average slope of its transfer function (see Fig. 7.13).

Fig. 7.13 Linearity of a dac

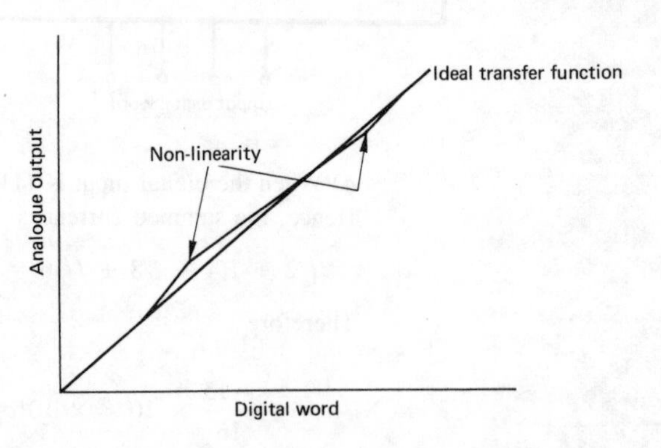

Monotonicity A dac is said to be monotonic if an increase in the input digital word never results in a decrease in the analogue output voltage, or vice versa.

Zero Offset The zero offset of a dac is the difference between the analogue output voltage when the digital input word signals zero output voltage, and the actual zero output voltage.

Maximum Throughput Rate This is the maximum number of digital-to-analogue conversions the circuit is able to perform per second.

Settling Time When the input digital word changes suddenly over a wide range, the output voltage will not be able to immediately attain its new value. Figure 7.14 shows how the output voltage will exhibit some overshoot and take some time to settle to its new value. The output voltage is said to have settled when it remains within a specified band (usually ±lsb/2) of values about its final value. The settling time is shown by the figure.

Fig. 7.14 Settling time of a dac

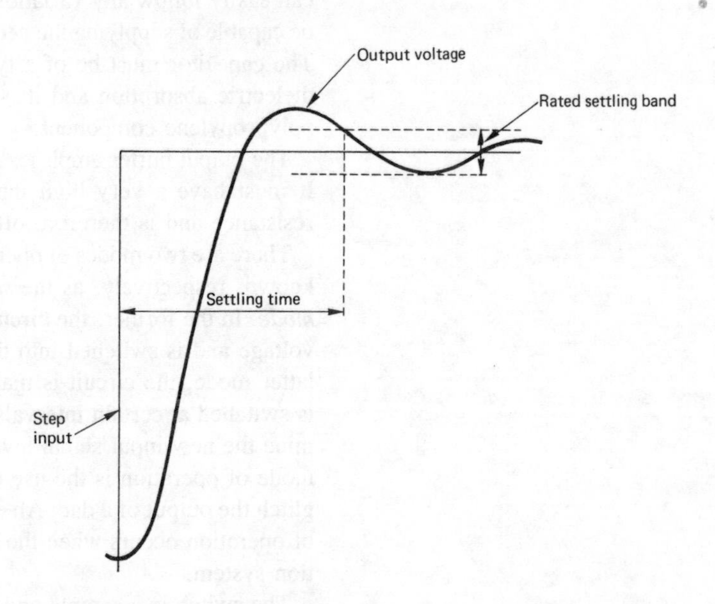

Sample-and-Hold Amplifiers

A **sample-and-hold amplifier** is a circuit that can sample an analogue voltage and then hold the instantaneous value upon the receipt of a logic command signal. Essentially, the circuit consists of a capacitor and an electronic switch, the capacitor acting as the storage element. The basic circuit of a sample-and-hold amplifier is shown by Fig. 7.15. When the switch is closed, the output voltage V_{out} will be equal to the input voltage V_{in}. When the switch is opened, the charge stored in the capacitor is retained, and so the output voltage is held at the value of the input voltage at the instant the switch was opened.

Fig. 7.15 Basic sample-and-hold circuit

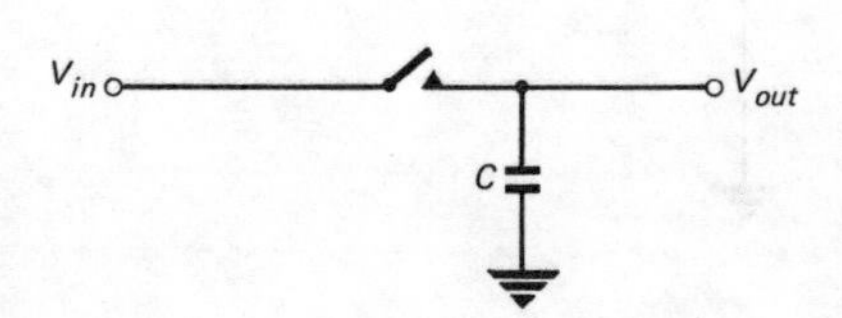

In order to prevent the storage capacitor loading the analogue source, an input unity-gain amplifier is normally employed. Also, in order to prevent the capacitor discharging rapidly into the load (which

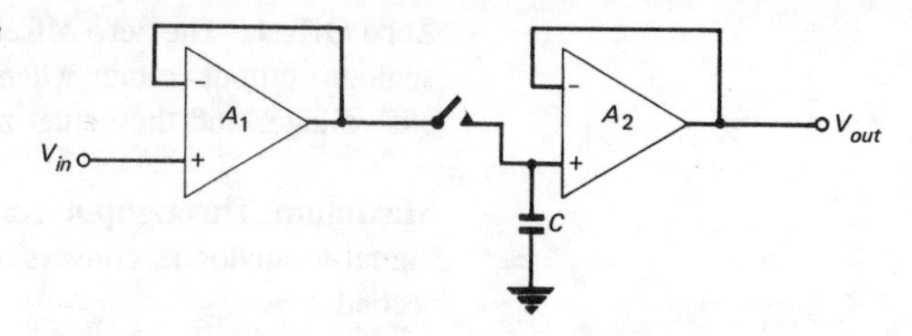

Fig. 7.16 Sample-and-hold amplifier

may well be of fairly low resistance), another unity-gain amplifier is placed at the output of the circuit. The basic circuit using two buffer amplifiers is given in Fig. 7.16.

The input buffer amplifier A_1 is an op-amp connected to operate as a voltage follower. It therefore has a high-input resistance and a low-output resistance. The high-input resistance ensures that the source is not loaded. The low-output resistance is required to keep the time constant small so that the voltage developed across the capacitor C can easily follow any variations in the input signal voltage. A_1 must be capable of supplying the necessary charge current to the capacitor. The capacitor must be of a type that has a low leakage and a low dielectric absorption and it is commonly either a polystyrene or a polypropylene component.

The output buffer amplifier A_2 is also an op-amp voltage follower. It must have a very high input resistance and a very low output resistance and is therefore often a fet input type (bifet or bimos).

There are two modes of operation for a sample-and-hold amplifier, known, respectively, as the *sample* or *tracking mode* and the *hold mode*. In the former, the circuit continuously tracks the input signal voltage and is switched into the hold mode at certain times. In the latter mode, the circuit is mainly holding the sampled voltage and is switched at certain intervals to sample just long enough to determine the new input signal level. An example of the use of the first mode of operation is the use of a sample-and-hold amplifier to deglitch the output of a dac. An example of the use of the second mode of operation occurs when the circuit is employed in a data acquisition system.

The switch in a sample-and-hold amplifier is usually provided by a fet and Fig. 7.17 gives a typical circuit. The switch is provided

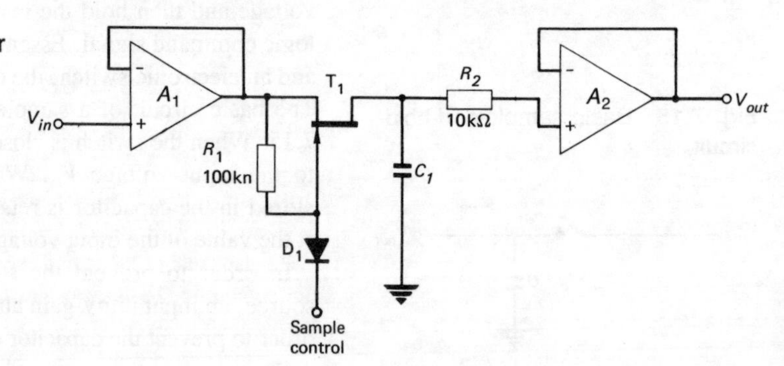

Fig. 7.17 Sample-and-hold amplifier

by the jfet T_1 and is turned ON or OFF by a voltage applied to the *sample control* input. When this voltage is high, T_1 turns ON and the voltage at the output of voltage follower A_1 is applied to the storage capacitor C_1. If the sample control voltage is low, T_1 turns OFF and the sampled input voltage is held by C_1. The charge on C_1 will slowly disappear because of the finite resistance of T_1 and the op-amp's bias current. Resistor R_2 is normally fitted merely as protection for the input stage of the op-amp. If the bias current of the op-amp is I_1, the jfet leakage current is I_2 then the *drift rate* is

$$dV/dt = (I_1 + I_2)/C_1$$

The time taken for the capacitor voltage to reach the same value as the input voltage depends upon the time constant of the circuit. This is equal to

$$\tau = (R_{DS(on)} + R_{out})C_1$$

The output resistance R_{out} of the voltage follower will be low and can usually be neglected. After a time equal to *five* time constants, the capacitor voltage will be within 1% of the input voltage. After 7τ the error will be only 0.1%.

Example 7.4

If the ON resistance of jfet T_1 in Fig. 7.17 is 30 Ω and $C_1 = 1\ \mu F$, calculate the time required for the capacitor voltage to reach a value within 1% of the input voltage.

Solution

Time constant $\tau = C_1 R_{DS(on)} = 1 \times 10^{-6} \times 30 = 30 \times 10^{-6}$ s
Therefore $5\tau = 150 \times 10^{-6}$ s *(Ans)*

Example 7.5

Calculate the maximum capacitor voltage that can be employed if $R_{DS(on)}$ = 35 Ω and if the maximum time for the output voltage to be within 1% of the output voltage is to be 10 μs. If the op-amp bias current is 150 pA and the fet leakage current is 200 pA, calculate the drift rate using this value of capacitance.

Solution

$$10 \times 10^{-6} = 5\ C_1 \times 35$$

Therefore $C_1 = \dfrac{10 \times 10^{-6}}{5 \times 35} = 0.057\ \mu F$ *(Ans)*

Drift rate $= dV/dt = I/C_1 = \dfrac{350 \times 10^{-12}}{0.057 \times 10^{-6}} = 6.14$ mV/s *(Ans)*

The drift rate of the stored charge can be considerably reduced if the circuit is modified in the way shown by Fig. 7.18. A description of its action is left as an exercise (7.9).

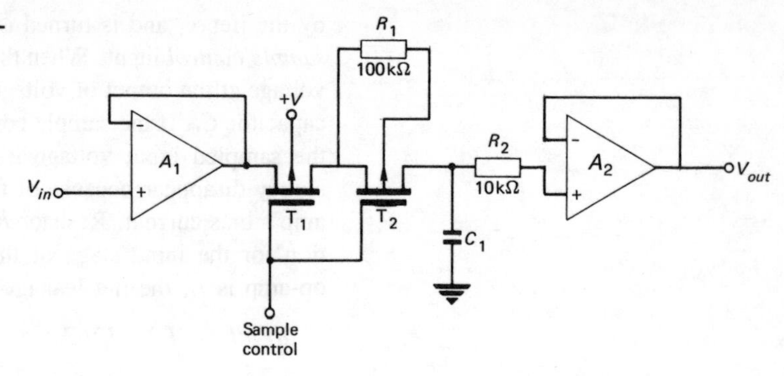

Fig. 7.18 Low-drift-rate sample-and-hold amplifier

Terms used with Sample-and-Hold Amplifiers

Acquisition Time The acquisition time of a sample-and-hold amplifier is the minimum time required for the output voltage to begin tracking the input voltage, within some specified accuracy, after the sample command has been applied (Fig. 7.19).

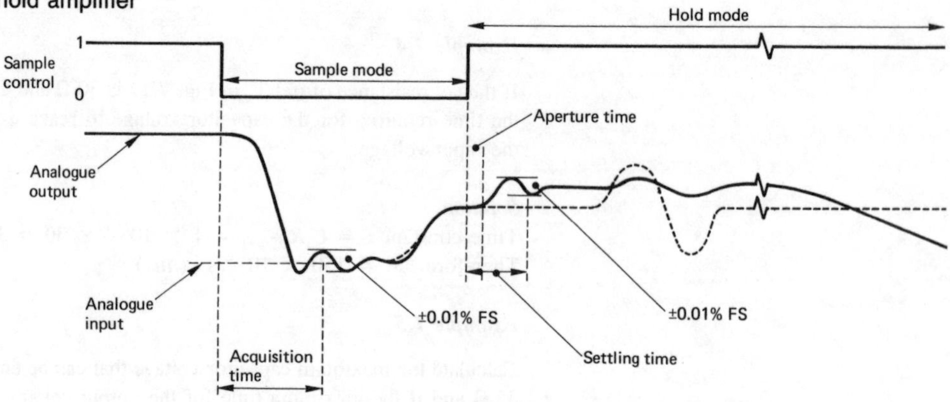

Fig. 7.19 Illustrating the acquisition time, aperture time, and settling time of a sample-and-hold amplifier

Aperture Time The aperture time of a sample-and-hold amplifier is the time that elapses between the application of the hold command and the output voltage ceasing to track the input voltage (Fig. 7.19).

Drift Rate or Droop Rate The drift or droop rate is the maximum rate of change of the output voltage when the circuit is in its hold mode.

Settling Time The settling time is the time required for the output voltage to track the input voltage within a certain error band (Fig. 7.19).

Hold Step This is the magnitude of the step in the output voltage produced when the circuit is switched from the sample mode into the hold mode (Fig. 7.19).

Analogue-to-Digital Converters

The function of an **analogue-to-digital converter** is to convert an input analogue signal voltage into the equivalent output digital codeword. A number of different circuit techniques are presently available and these are, in the main, some version either of a feed-back technique or of an integrating technique.

If the adc is to convert the input signal without error, the input signal must not change during the conversion time. For this reason the adc is often preceded by a sample-and-hold amplifier. The timing diagram of a sh/adc system is shown by Fig. 7.20 which supposes that the conversion time is equal to the time occupied by 16 clock pulses.

Fig. 7.20 Timing diagram of a sample-and-hold amplifier

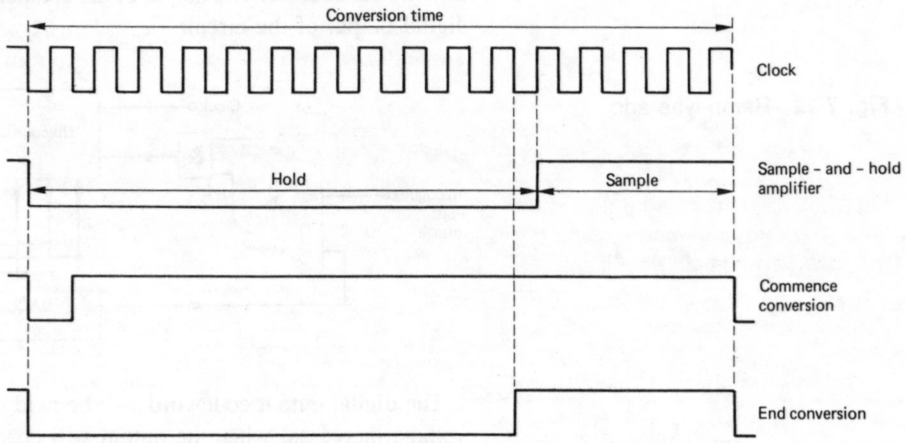

Fig. 7.21 Transfer function of a 3-bit adc

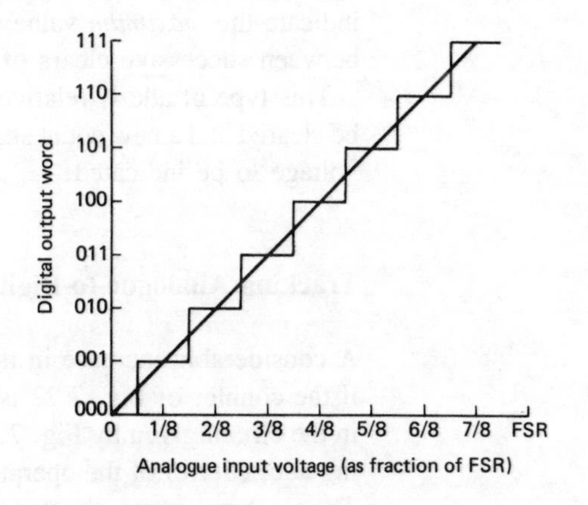

Figure 7.21 shows the theoretical transfer function of a 3-bit adc. A step characteristic is obtained because a range of input analogue voltages ± lsb either side of a quantization level have the same digital codeword. The ideal transfer characteristic is the straight line drawn between the points 0,000 and 7/8FSR,111.

Ramp Analogue-to-Digital Converter

The basic circuit of a ramp-type adc is shown by Fig. 7.22. At the start of a conversion, the counter is reset to a count of 0. Clock pulses are applied to the counter and they cause the parallel counter outputs to act accordingly. The counter outputs are applied to the dac as shown. As a result, the output of the dac is increased in a number of steps, each of which is equal to one lsb. The dac output is fed to the input of an op-amp connected so as to operate as a voltage comparator. When the output of the dac becomes equal to the analogue input signal voltage, the output of the comparator changes state and disables the counter. The output of the counter then gives the required digital output of the circuit.

Fig. 7.22 Ramp-type adc

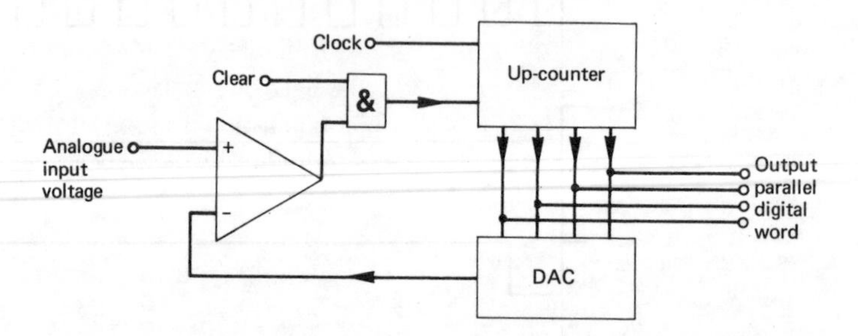

The digital output codeword will be held until the analogue input voltage increases, when the output will change to indicate the new appropriate value. The digital output of the circuit therefore can only indicate the *maximum* value of the analogue input voltage attained between successive clears of the adc.

This type of adc is relatively slow in its operation since it has to be cleared and a new count started from zero for any decrease in input voltage to be indicated.

Tracking Analogue-to-Digital Converter

A considerable increase in the speed of operation can be achieved if the counter of Fig. 7.22 is replaced with an up-down counter as in the circuit given by Fig. 7.23. In this circuit, known as the tracking adc, control of the operation is vested in the up–down counter. The clock pulses are directed to either the up-count or the down-count input of the counter depending on whether or not the dac output is respectively larger or smaller than the analogue input voltage. If the dac output is smaller than the analogue input voltage, the comparator output will be high. Then all inputs to the top gate are high so that its output is also high. The lower gate then has one of its inputs low so that its output is low. The counter is then in its up-count mode of operation.

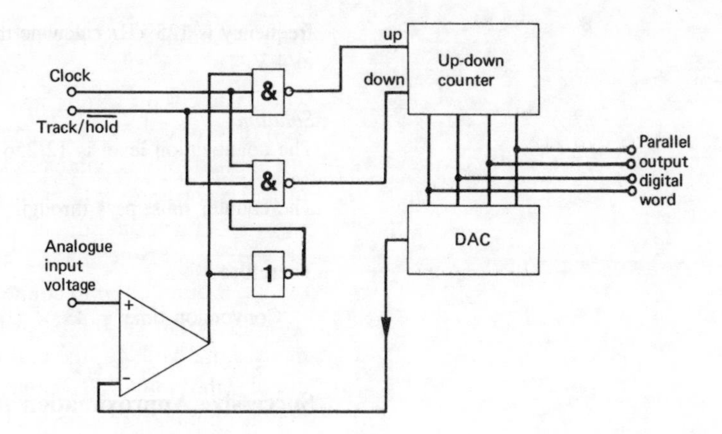

Fig. 7.23 Tracking adc

If the dac output becomes larger than the input analogue voltage, the gate outputs will switch states and then the counter will go into its count-down mode of operation. The tracking adc is thus able to continuously track any variations in the input analogue voltage and maintain an updated digital output codeword. If required, the circuit can be caused to hold at a particular digital output by taking the track/$\overline{\text{hold}}$ input terminal low.

Example 7.6

An 8-bit counter ramp adc has a 100 kHz clock frequency. Calculate its maximum conversion time for a full-amplitude range input signal.

Solution

For an 8-bit converter there are $2^8 = 256$ different codewords. This means that the maximum conversion time is 256 clock periods or

$$256 \times 1/10^5 = 2.56 \text{ ms} \quad (Ans)$$

Example 7.7

An 8-bit counter ramp adc has a maximum range input voltage of +12 V. If the clock frequency is 125 kHz determine the time taken to convert a 6 V input signal voltage.

Solution

8 bits produce 256 codewords. Therefore, the quantization interval is 12/256 = 46.88 mV. The counter must ramp through

$$6000/46.88 = 128 \text{ codewords.}$$

Therefore

$$\text{Conversion time} = 128 \times (1/125) \times 10^{-3} = 1.024 \text{ ms} \quad (Ans)$$

Example 7.8

A tracking-type adc has a maximum range input voltage of 12 V. If the clock frequency is 125 kHz calculate the conversion time if a 6 V input changes to 4 V.

Solution

The quantization level is 12/256 = 46.88 mV.

The counter must pass through $\frac{6000 - 4000}{46.88} \simeq 43$ codewords.

Therefore

$$\text{Conversion time} = 43 \times (1/125) \times 10^{-3} = 0.344 \text{ ms} \quad (\textit{Ans})$$

Successive Approximation ADC

The block diagram of a **successive approximation adc** is shown in Fig. 7.24. It is probably the most popular kind of adc for moderate-to high-speed conversion applications since it is faster than either of the two previous types.

The dac is controlled by the shift register in such a way that a conversion is completed in only n steps, where n is the number of bits held by the register. At the beginning of a conversion, the register is set to zero. The next clock pulse applies the msb of the register output to the dac and its *half-full scale range* (1/2 FSR) analogue output is applied to the voltage comparator.

Fig. 7.24 Successive approximation adc

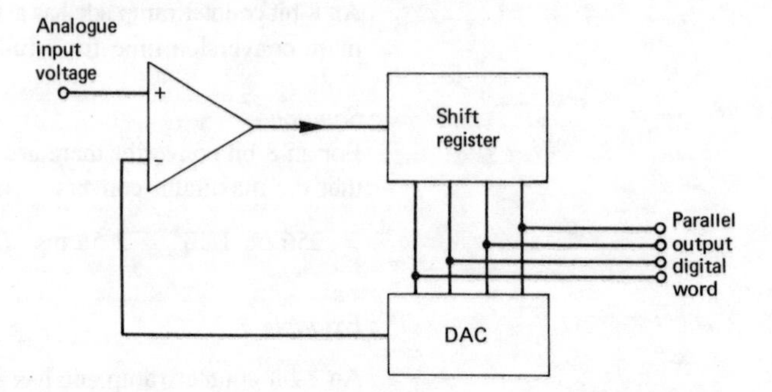

If the dac output is smaller than the analogue input signal voltage, the msb output of the register is retained. Conversely, if the dac output is greater than the input voltage, the msb output is turned OFF. The next clock pulse turns ON the next msb output of the register and this causes the dac to generate a voltage of 3/4 FSR if the msb is still ON or 1/4 FSR if the msb is OFF. The new dac output is compared with the input voltage and a decision made either to retain the next msb or to turn it OFF. The process of turning ON a register output and then comparing the total dac output with the analogue input voltage continues until the lsb is reached. Thus, after n comparisons the outputs of the shift register produce the required digital output codeword.

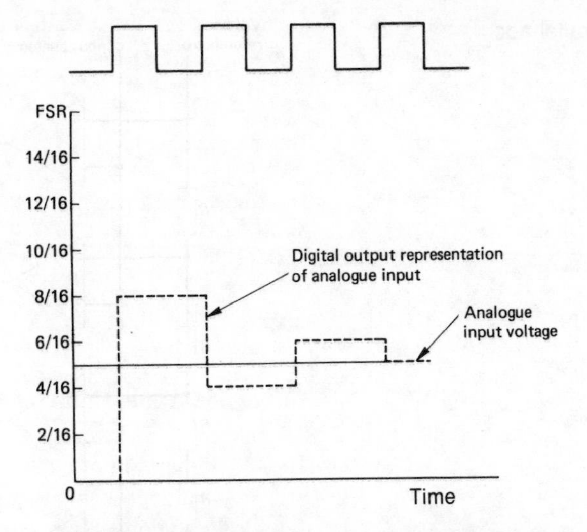

Fig. 7.25 Action of a successive approximation adc

The approximation process is illustrated by Fig. 7.25 for a 4-bit adc in which the analogue voltage is equal to 5/16 FSR.

The first approximation is 8/16 FSR. This is too big and so the next clock pulse turns the msb OFF and the second approximation is 4/16 FSR. This value is less than the output voltage and so this bit is retained. The third clock pulse turns ON the next register bit and the dac output is incremented by 2/16 FSR to 6/16 FSR. Now the approximation is again too large and so the next clock pulse turns the third bit OFF and the lsb ON. The dac output is then (4/16 FSR + 1/16 FSR) or 5/16 FSR, and this is equal to the analogue input voltage. The codeword produced by the dac is then 0101.

Example 7.9

Calculate the percentage increase in conversion speed if an 8-bit successive approximation adc is employed instead of an 8-bit ramp adc. Assume the maximum input voltage is applied to both circuits.

Solution

For an 8-bit ramp adc, the circuit must pass through $2^8 = 256$ states. Hence the conversion time is equal to 256 times the periodic time T of the clock.

The successive approximation adc will carry out the conversion in $8 + 1 = 9$ clock periods.

Therefore, increase in conversion time is

$$\frac{256 - 9}{9} \times 100\% = 2744\% \quad (Ans)$$

Parallel or Flash ADC

When very fast analogue-to-digital conversion is required the **parallel** or **flash** circuit is often employed. Figure 7.26 shows the circuit of an

Fig. 7.26 Parallel adc

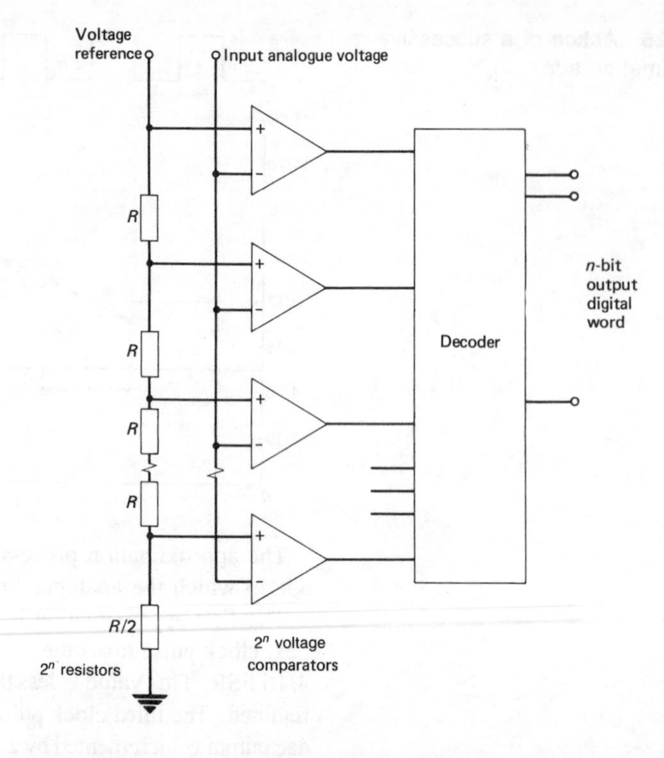

n-bit parallel adc. The circuit has 2^n resistors connected in series between a positive voltage reference and earth. Also, 2^n voltage comparators are connected to the junctions of these resistors as shown, so that the comparator inputs are spaced at voltage intervals equal to the lsb. The analogue input signal level is simultaneously compared against 2^n d.c. voltage levels. All those comparators which are biased to a voltage smaller than the analogue voltage turn ON, but all those comparators biased above the input voltage remain OFF. The outputs of the comparators are then decoded to give the required output digital codeword.

Suppose that the voltage reference is 6 V and that a 4-bit circuit is employed. There are then 2^4 (or 16) series resistors and 16 voltage comparators. The voltages in the resistor chain will increment in 1 V steps and hence only the bottom 6 comparators will turn ON. The outputs of these six comparators are then decoded to give the output codeword 0110.

The parallel adc is capable of very high speed conversions but it suffers from the disadvantage that even low-resolution circuits need a large number of voltage comparators.

Integrating-type ADC

A number of adc circuits operate by converting the analogue input voltage into a proportional time interval and then, in some way,

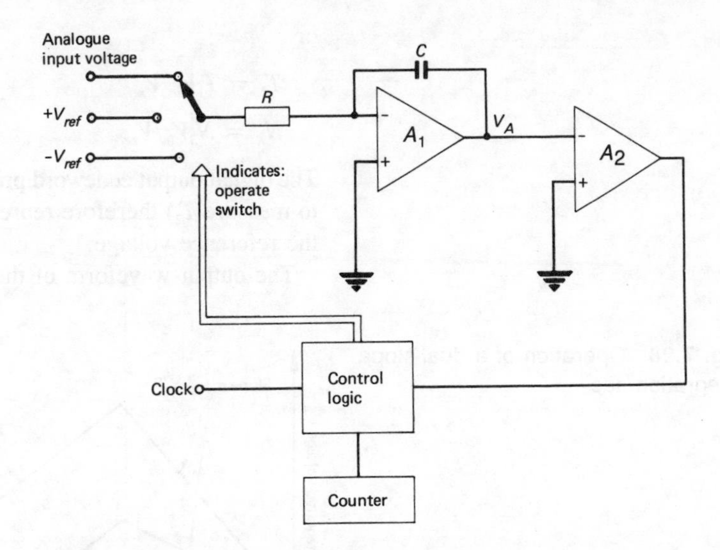

Fig. 7.27 Dual-slope integration adc

measuring its duration. The circuit of a *dual-slope integration adc* is shown by Fig. 7.27.

The input analogue signal is applied to the integrator (previously set to output zero) and is integrated for a fixed time T_1. The output voltage V_A of the integrator is then

$$V_A = \frac{1}{RC}\int_0^{T_1} V_{in}\mathrm{d}t$$

If the input voltage were constant over the time period T_1, the output voltage V_A would be a linear ramp waveform. Then at $t = T_1$,

$$V_A = \frac{V}{RC}T_1$$

During this time, the counter has counted N_S clock pulses. At the end of T_1 seconds, the control logic switches the integrator to either the $+V_{ref}$ or the $-V_{ref}$ input, whichever is of the opposite polarity to the input voltage. This makes the integrating capacitor discharge and consequently V_A falls linearly towards zero volts. Clock pulses are again counted whilst V_A falls. When V_A becomes equal to 0 V, the comparator changes state and stops the counter at a count of N_R pulses. The time T_2 taken for V_A to fall to zero is directly proportional to the average value of the analogue input voltage. Therefore,

$$V_A = \frac{1}{RC}V_{ref}T_2$$

and

$$\frac{V_{ref}T_2}{RC} = \frac{V_{in}T_1}{RC}$$

or

$$T_2 = T_1 V_{in}/V_{ref}$$

$$N_R = N_S V_{in}/V_{ref}$$

The digital output codeword produced by the counter (counting pulses to measure T_2) therefore represents the ratio of the input voltage to the reference voltage.

The output waveform of the integrator is shown by Fig. 7.28.

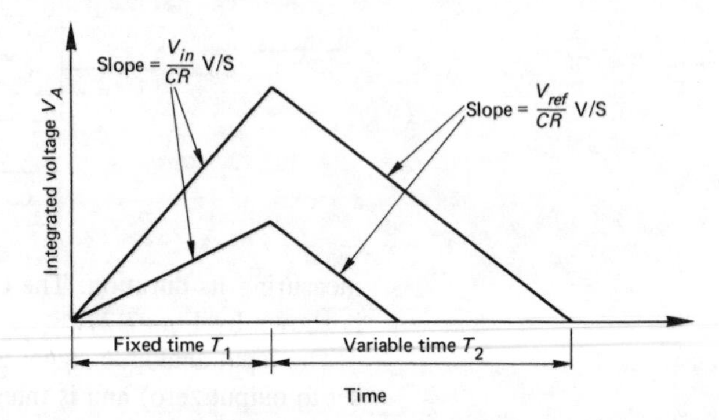

Fig. 7.28 Operation of a dual-slope integration adc

The dual-slope technique of analogue-to-digital conversion possesses some important advantages: *a*) The accuracy of a conversion is independent of the stability of both the clock and the integrator provided they are constant during the conversion period. *b*) the noise immunity can be quite high.

Most commercial adcs are either of the successive approximation type, or of the dual-slope integration type. The former is employed in high-speed, and the latter in low-speed, applications.

Terms used with ADCs

Absolute Accuracy The absolute accuracy of an adc is a measure of the circuit's departure from the ideal transfer characteristic, i.e. it is the difference between the theoretical and actual inputs needed to produce a given output codeword.

Relative Accuracy The relative accuracy of an adc is a measure of the largest deviation between the circuit's transfer function and a straight line drawn through the point 0 V and the FSR. It is expressed as a percentage of the FSR.

Differential Linearity and Monotonicity See digital-to-analogue converters (p. 192).

Dynamic Range This is the ratio of the full-scale range FSR to the smallest change in the input voltage the circuit is able to register.

Resolution The resolution of an adc is the smallest change in the input analogue voltage that can be indicated at the output of the circuit.

Data Acquisition Systems

A **data acquisition system** provides an interface between the world of physical parameters, which are all analogue in their nature, and the digital world in which various computation and control exercises on the parameters are to be carried out.

Fig. 7.29 Data Acquisition System

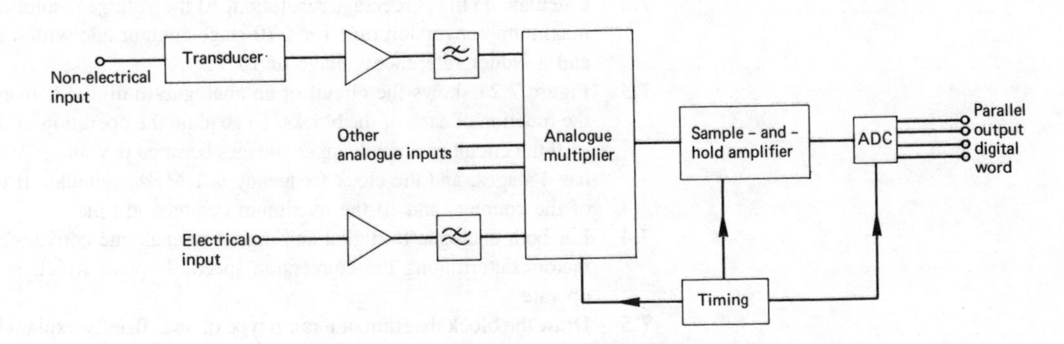

Essentially, a data acquisition system consists of an analogue multiplexer, a sample-and-hold amplifier, and an adc. The basic block diagram of such a system is shown in Fig. 7.29. The system has a number of analogue signal inputs, some of which are electrical in nature but others are not. These latter, which may for example be temperature, position, flow, acceleration or pressure data, are first converted into electrical signals by means of a *transducer*. The electrical signals are first amplified and are then filtered to remove noise and any other unwanted components that may be present. The analogue signals originating from a number of different sources are applied to an *analogue multiplexer*. This circuit is similar in concept to the digital multiplexer discussed in Chapter 3 and its function is to sequentially switch its inputs to its single output terminal. Thus, each analogue input is, in turn, connected for a short period of time to the sample-and-hold amplifier. This circuit holds the multiplexed voltages while the adc changes them into the corresponding digital form. The digital output is then passed to the input of, usually, a digital computer or a microprocessor.

A number of variants on the basic theme are equally possible.

1 Low-level multiplexing may be employed in which the amplifier follows the multiplexer; this arrangement demands the provision of only one amplifier but it may be necessary for the amplifier to be able to vary its gain for each channel.

2 The analogue signals may be converted to digital form at each of the transducer outputs and the digital data may then be sent in serial form to the point of processing. There it can be changed into parallel form and then multiplexed with the other channels.

3 Figure 7.29 can be modified so that there is a sample-and-hold amplifier in each channel prior to the multiplexer.

Exercises 7

7.1 Explain the operation of a summing amplifier type of digital-to-analogue converter. List, and briefly explain, its sources of error.

7.2 Calculate *a*) the percentage resolution, *b*) the voltage resolution, and *c*) the maximum conversion time for a 10-stage counter adc with a 1 MHz clock and a ladder reference voltage of 16 V.

7.3 Figure 7.24 shows the circuit of an analogue-to-digital converter. *a*) State the function of each of the blocks. *b*) Explain the operation of the converter. *c*) If the circuit can handle input voltages between 0 V and +5 V, the counter has 4 stages, and the clock frequency is 1 MHz, calculate i) the resolution of the counter, and ii) the maximum conversion time.

7.4 For both analogue-to-digital and digital-to-analogue converters explain the factors determining the conversion speed. Explain which is the faster to operate.

7.5 Draw the block diagram of a ramp type of adc. Briefly explain its operation.
Calculate the number of stages required in the dac in order for the adc to have a resolution of 10 mV when the voltage reference is 10 V.

7.6 For the adc in **7.5** calculate the clock frequency necessary if the circuit is to perform 100 conversions per second.

7.7 Explain the operation of a ramp-type dac. List, and briefly explain, its main sources of error.

7.8 Describe the operation of the circuit given in Fig. 7.18.

Index